TENDERS AND CONTRACTS
FOR BUILDING

THIRD EDITION

Other titles by the Aqua Group

Pre-Contract Practice for the Building Team

Contract Administration for the Building Team

TENDERS AND CONTRACTS FOR BUILDING

THIRD EDITION

THE AQUA GROUP

With sketches by
Brian Bagnall

**Blackwell
Science**

© The Aqua Group 1975, 1982, 1990, 1999

Blackwell Science Ltd
Editorial Offices:
Osney Mead, Oxford OX2 0EL
25 John Street, London WC1N 2BL
23 Ainslie Place, Edinburgh EH3 6AJ
350 Main Street, Malden
    MA 02148 5018, USA
54 University Street, Carlton
    Victoria 3053, Australia
10, rue Casimir Delavigne
    75006 Paris, France

Other Editorial Offices:

Blackwell Wissenschafts-Verlag GmbH
    Kurfürstendamm 57
    10707 Berlin, Germany

Blackwell Science KK
MG Kodenmacho Building
7–10 Kodenmacho Nihombashi
Chuo-ku, Tokyo 104, Japan

First published in 1975 by Crosby
    Lockwood Staples (under the title 'Which
    Builder?'; original ISBN 0 258 97041 3)
Published in 1982 by Granada Publishing
    (under the title 'Tenders and Contracts
    for Building'; ISBN 0 246 118385)
Reprinted 1984
Reprinted by Collins Professional and
    Technical Books 1986
Second edition published by Blackwell
    Scientific Publications 1990
Reprinted 1992
Reprinted 1996 by Blackwell Science
Third edition 1999

Set in 10/12.5 pt Century
    DP Photosetting, Aylesbury, Bucks
    ed and bound in Great Britain by
    Books Ltd, Bodmin, Cornwall

        well Science logo is a trade mark
            Science Ltd, registered at the
            m Trade Marks Registry

DISTRIBUTORS

Marston Book Services Ltd
PO Box 269
Abingdon
Oxford OX14 4YN
(*Orders:* Tel: 01235 465500
        Fax: 01235 465555)

USA
Blackwell Science, Inc.
Commerce Place
350 Main Street
Malden, MA 02148 5018
(*Orders:* Tel: 800 759 6102
        781 388 8250
    Fax: 781 388 8255)

Canada
Login Brothers Book Company
324 Saulteaux Crescent
Winnipeg, Manitoba R3J 3T2
(*Orders:* Tel: 204 837-2987
        Fax: 204 837-3116)

Australia
Blackwell Science Pty Ltd
54 University Street
Carlton, Victoria 3053
(*Orders:* Tel: 03 9347 0300
        Fax: 03 9347 5001)

A catalogue record for this title is
available from the British Library

ISBN 0–632–04277–X

Library of Congress
Cataloging-in-Publication Data is available

For further information on
Blackwell Science, visit our website:
www.blackwell-science.com

Contents

Introduction

Although *Tenders and Contracts* was the last of our trilogy of texts to appear, it now forms the starting point for our examination of the whole subject of procurement. It is significant that, although this is only the third edition (or fourth if you include the original title *Which Builder?*) and *Pre-Contract Practice* and *Contract Administration* are in their eighth editions, *Tenders and Contracts* is perhaps the most important. *Tenders and Contracts* examines the broader aspects of the subject of procurement, and identifies the criteria that need to be addressed before the choice of a specific tendering procedure and contractual arrangement is made. Once that choice is made, the detailed considerations set out in *Pre-Contract Practice* and *Contract Administration* come into play.

It now almost goes without saying that the increasing rate of change in the construction industry requires the constant updating of our approach to procurement – but such is the case. Indeed, the drivers of change have included some very influential contributors of late.

Sir Michael Latham's report, *Constructing the Team*, has been followed by Sir John Egan's report *Rethinking Construction*, the Reading Construction Forum's publication *Trusting the team: the best practice guide to partnering in construction* and the European Construction Institute's document entitled *Partnering in the public sector: a toolkit for the implementation of post award, project specific, partnering on construction projects*. Government strategy has also changed with the introduction of the private finance initiative (PFI).

In addition, there has been some significant legislation, especially the Construction (Design and Management) Regulations 1994 and the Housing Grants, Construction and Regeneration Act 1996, both of which have had a significant effect on the industry – the implifications of which are discussed in this new edition.

While examining these latest influences, we have maintained the original essence of the book – an explanation of the basic concepts of the different approaches to procurement. Although extending, and in some cases rewriting chapters, we have not, on this occasion, altered the running order. This is not to say that we have not considered the matter – our conclusion, however, is that the most logical order of the chapters varies, depending on the reader's situation or starting point and on their priorities and the perceived options open to them. We venture to suggest, therefore, that this book is not a 'straight read', but should be read selectively according to the individual's needs. A useful Glossary is included at the end.

Chapters 1 and 2 have been rewritten, and retitled, to provide a clearer up-to-date analysis of the principles of procurement, and to review the criteria to be considered in setting a procurement strategy. Chapter 1 includes references to the Construction (Design and Management) Regulations 1994 (CDM), and considers risk and accountability, while Chapter 2 covers the concepts of partnering, value management, the private finance initiative and the Housing Grants, Construction and Regeneration Act 1996.

Chapter 5 has also been rewritten and retitled to give a better and fuller appreciation of the range of documentation available to support different methods of procurement. The practical uses of the various documents, and sometimes more importantly, their limitations, are identified, as well as their contractual significance.

Chapter 11 considers management and construction management contracts, with a subtle change of title to reflect the change in emphasis in this section of the market. This chapter has been considerably extended to include a more detailed assessment of the advantages and disadvantages of the system.

Chapter 12, on design and build contracts, has been extensively revised and extended to provide a more detailed examination of the system, which continues to occupy a prominent position in the industry. A new section, on managing the design process, highlights the problems of novation and an explanation is given of the significant differences in the JCT Design and Build form of contract, compared with the usual consultant design versions.

Chapter 14 on partnering, is entirely new and reflects the significance attached to recent developments emanating from Sir Michael Latham's report, *Constructing the Team* referred to above. This is seen to be an important area of development for the future, especially for those clients able to take advantage of continuity.

All other chapters have been extended and updated and more diagrams have been included to add a visual dimension to the understanding of the comparative merits of the different systems.

Our appreciation is due to John Townsend, who undertook the task of the final editing of this edition and who put together most of the rewritten chapters – John is the only current full time member of the Aqua Group who was involved in its first edition, under the title of *Which Builder?* We must not forget, however, that Brian Bagnall is also still involved in adding his inimitable cartoons to the text – these add the icing to the cake, which we hope our readers will enjoy. As always, we are grateful to Julia Burden, our Commissioning Editor, for her assistance and encouragement.

The Aqua Group:

BRIAN BAGNALL BArch(Liverpool)
OLIVER BURSTON AADipl, RIBA
JOHN CAVILLA BSc(Hons), MCIOB, MIMgt, MAPM
RICHARD OAKES BSc(Hons), FRICS
QUENTIN PICKARD BA, RIBA (Chairman)
GEOFFREY POOLE FRIBA, ACIArb
JOHN TOWNSEND DipArb, FRICS, FCIArb
JOHN WILLCOCK DipArch, RIBA
JAMES WILLIAMS DA(Edin), RIBA

'... proper time ...'

Chapter 1
Principles of Procurement

Simple theory – complex practice

Procurement, the process of obtaining goods and services from another for some consideration, is one of the oldest transactions known to man. It can also be one of the simplest. From time immemorial, man has been unable, or unwilling, to make himself totally self-sufficient, and has had to trade with his neighbour. The earliest deals were struck by barter, one to one, face to face, with immediate effect. Nothing could be simpler. Some are still struck on that basis, but not many in construction. Today's construction deals are often vast in scope, involve several designers, require specialist contractors, are executed at arm's length and take considerable time to complete.

This book examines the types of contract conditions which are appropriate in any given circumstances and the various ways in which the consideration may be arrived at.

However, complex though modern construction projects may be, they are still based on a number of comparatively simple principles – cost, quality and time. The complexity arises from the need to meet numerous regulations, to involve professional consultants, to achieve value for money, to demonstrate accountability, to regulate complicated contractual relationships and by the need to meet clients' required timescales.

The eternal triangle

From the client's point of view, there are probably only three criteria on his agenda. These are COST, QUALITY and TIME – he wants the highest quality, at the lowest cost, in the shortest time. Unfortunately, this is not always possible and a compromise has to be sought, based on the client's priorities. These criteria, and the compromise, can be visualised as a triangle (see Figure 1).

Figure 1. The eternal triangle.

If these three factors are kept in balance, with appropriate quality being achieved, at an acceptable price, in a reasonable timescale, the triangle would appear equilateral – with equal weight or emphasis being given to each factor. If, however, particular circumstances dictate that one of the factors must take precedence, the other two will 'suffer', or carry less weight or emphasis. The decision must lie with the client.

If quality is of paramount importance, adequate time must be allowed for the design and specification to be perfected and cost will rise on both counts. If speed of completion is paramount, quality and cost may both have to suffer. If lowest cost is the priority, time may not be prejudiced, but quality will suffer. Again, the weight or emphasis of these criteria can be shown diagramatically (see Figure 2).

Figure 2. Variations on the eternal triangle, showing different priorities.

The effect of these different priorities is relative, and there is no reason why, with proper planning and management, those elements with a lower priority cannot be adequately controlled. By evaluating the factors discussed in this and subsequent chapters, it is anticipated that the design team will be able to devise a tendering procedure and contractual arrangement that will obtain the best value for money in the circumstances – indeed, this is the objective of this book.

It is therefore of vital importance that the strategy for the client's construction project is discussed at the earliest opportunity, to achieve an appropriate procurement system, tuned to the client's requirements and priorities.

Other considerations

However, these are not the only criteria to be discussed at this stage. There are others, which the client may not immediately be aware of, and which may affect the decision on the most suitable procurement system, such as:

- CDM (the Construction (Design and Management) Regulations)
- risk
- accountability.

CDM – the Construction (Design and Management) Regulations 1994

The Construction (Design and Management) Regulations 1994, which came into effect on 31 March 1995, place statutory duties on clients, planning supervisors, designers and contractors, to plan, co-ordinate and manage health and safety throughout all stages of a construction project.

Under the Regulations, the employer is charged with appointing a planning supervisor, who must be competent, and giving him the relevant information. The employer must also ensure that the designers he appoints are competent in terms of health and safety, and that they make adequate provisions for health and safety.

Although the Regulations introduce no new laws on health and safety, they codify the administrative machinery required for the management of health and safety and provide an audit trail to demonstrate accountability. What the Regulations do introduce, however, is the sanction of a criminal penalty in the event of default. This is therefore a serious matter, and the requirements of the Regulations must be identified at the very earliest stage in the strategic planning process.

Risk

Risk is defined and explained in detail in Chapter 2. At this stage, it is sufficient to note that decisions on the allocation of risk between the client and the contractor can have a fundamental effect on the choice of the right tendering procedure and contractual arrangement.

It is therefore important to establish the client's attitude to risk at the earliest opportunity. If the client is prepared to accept increased risk, the consultants will have much greater flexibility to devise an effective strategy based on non-traditional procedures.

Accountability

Accountability is closely linked with risk and is considered in Chapter 3. However, decisions at an early stage can again have a significant effect on tendering procedures and contractual arrangements. It is therefore essential that the client identifies the level of accountability required on the project as early as possible, and for the consultants to explain the principles involved, so that an early decision can be made.

Making the contract

The process of making the contract may be divided into three parts:

- deciding on the type of contract and the particular terms and conditions under which the work will be carried out
- selecting the contractor
- establishing the contract price, or how the price will be arrived at.

Type of contract

The choice of the type of contract, and the particular terms and conditions under which the work will be carried out, will normally be made by the client in the light of advice he receives from his professional advisers. The choice must be made at an early stage, as it will affect the way in which the contract documentation is prepared. If competitive tendering is to be involved, the type of contract and the actual conditions to be used must be defined before contractors are invited to tender. If negotiation with a single contractor is contemplated, the final decision may be made in consultation with him, after he has been appointed. In traditional single stage competitive tendering, the method of establishing the contract price must also be decided before tenders are invited. In negotiated contracts, this decision can also be delayed until the contractor has been appointed.

The range of contracts available to choose from is considerable, and within each general type of contract a choice can be made as to the particular terms and conditions which are most suitable to the circumstances. A list of contracts published by the Joint Contracts Tribunal (JCT), the Association of Consulting Architects (ACA) and the Government are set out in Appendix B, and a brief summary is set out below.

(1) fixed price contracts, with or without fluctuations, based on:
       performance specification

      specification and drawings
      schedule of rates
      bills of approximate quantities
      bills of quantities

(2)   cost reimbursement contracts
      prime cost + percentage fee
      prime cost + fixed fee

(3)   target cost contracts

(4)   management and construction management contracts

(5)   design/build contracts

(6)   continuity contracts.

Selection of the contractor – the tendering procedure

The selection of the contractor and the establishment of the contract price, or how the price will be arrived at, may be combined in a single operation or they may be two separate operations.

First it is helpful to look at a definition of tendering that has held good for a number of years:

The purpose of any tendering procedure is to select a suitable contractor, at a time appropriate to the circumstances, and to obtain from him at the proper time an acceptable tender or offer upon which a contract can be let.

In fact, this is not so much a definition as a statement of the purpose of tendering. However, it needs amplification for a full understanding of all that it covers. The statement emphasises that the purpose of tendering is the selection of the contractor and the obtaining of the tender or offer.

Although any tender must take account of the conditions of contract under which the work will be performed, the tendering procedure is not dependent on the type of contract to be used. The contract conditions have the same significance as, for example, contract drawings and bills of quantities, which together make up the total contractual arrangements. It is important to appreciate this, otherwise considerable confusion will prevail. Contractual arrangements are concerned with the type of contract to be entered into and the obligations, rights and liabilities of the parties to the contract. These may vary because of the type of project, but there is no direct relationship between them and the tendering procedure.

For example, the same contractual arrangements could prevail in two cases where, in one, the contractor had been selected after submission of a competitive tender based on detailed drawings and bills of quantities, while in the other, he had been selected on the basis of a business relationship with the client and single tender negotiations.

Having identified the differences between tendering procedures and contractual arrangements, let us consider in more detail the definition of tendering procedures. One of the first things to note is that the word 'time' occurs twice. On the first occasion it refers to the 'time appropriate' to select a suitable contractor. The second reference is to the 'proper time' to obtain a tender. In other words, the time for selection of the contractor may well be different from the time at which one obtains a tender or establishes the contract price. A further point to note is that the acceptable tender or offer does not finish the matter, but is the basis upon which a contract can be let.

It may be thought that an apparently simple definition has been examined in rather too much detail. This is deliberate and the succeeding chapters will give consideration to the various facets of this definition.

Establishing price and time

There has, in the past, been an underlying assumption that the tendering process was only concerned with finding out which contractor could submit the lowest price, based on the design presented to him. Only in recent years has greater consideration been given to situations where the contractor is partly or wholly involved in the design; where circumstances require that construction commences before the design has been completed; or where, for any other reason, price and time alone may not be the only factors to be taken into account.

In considering the purpose of tendering, one must think in the broadest terms. A tender quotes not just a price but also a standard of quality and a time in which to complete the work. The purpose of tendering may not be for one job; it may be necessary to consider it in terms of a total programme, of which that job is just one project.

The dynamics of tendering

Tendering procedure today is a dynamic situation. It is not a procedure which can be applied across the board on all construction works; in fact, 'procedure' is probably the wrong word, but, as it is used so frequently, it may be misleading not to use it here. There are so many different factors to consider in each case, and hence decisions to be made, that it would be more appropriate to refer to it as the tendering structure and requirement.

Much of what has been written with regard to past attitudes to tendering procedures also applies to contractual arrangements. There has been great emphasis on fixed price contracts with a somewhat poor realisation as to what the word 'fixed' entailed. Frequently the assumption has been that the contract should be a fixed price one, unless the situation is such as to make this impossible.

Although the actual form and the conditions of contract are part of the contractual arrangements, this book is not concerned with the legal contract as such. Numerous standard forms exist and a list of those in common use is given in Appendix B. The choice of the actual form to be used on any given project is a matter to be decided in relation to the nature of the work and the needs of the client, and is not a matter which will be considered in this book. Our concern relates to the general contractual arrangements, whether on a fixed price, cost reimbursement or target cost basis.

Finally, although everything written here is in terms of a contract between the client and the main contractor, it is equally applicable to a contract between a main contractor and a sub-contractor.

'... current trends ...'

Chapter 2
Basic Concepts and Current Trends

Chapter 1 dealt with the principles of procurement and the aims and purpose of tendering and contractual arrangements. This chapter deals with some of the basic concepts to be considered in selecting a contractor and formulating arrangements whereby a satisfactory contract may be made, and looks at current trends in the industry. New legislation and new policies on government procurement, particularly in relation to funding arrangements, necessitate a constant review of existing practices and the development of new ones.

Sir Michael Latham's report *Constructing the Team* initiated a review of many of the traditional practices in the construction industry. It highlighted the problems arising from processes that are uncompetitive, inefficient and expensive, from relationships that are ineffective, and from methods of working that are adversarial. Add to this the problems of skill shortages, inadequate training and insufficient investment, and it is easy to understand why the industry has such a poor image. Whether the industry will achieve the 30% cost savings, targeted by Sir Michael, remains to be seen, but there is plenty of scope for improvement – it is hoped that the matters discussed in this book will help.

In Sir John Egan's report *Rethinking Construction*, resulting from a government construction taskforce, the cost reductions sought have been raised to 40%. While praising the industry for engineering ingenuity, design flair and flexibility, the report stresses the need for improvement through product development, in project implementation and by partnering throughout the supply chain. Partnering is referred to later in this chapter and considered in greater detail in Chapter 14.

Traditionally, time and money were the only considerations in selecting a contractor and making a contract with him. Whilst this is true in the broadest terms, there are many factors which affect either time or the financial outcome of the contract, but which are not, at first, apparent as being directly related to either.

Although not every project will offer opportunities for the introduction of new techniques in terms of procurement procedure, the design team ought to review the possibility. Matters worth consideration, include

- the economic use of resources – labour, materials, plant and capital
- the contractor's contribution to design and contract programme
- production cost savings
- continuity
- risk and accountability.

Economic use of resources

It may be thought that the money paid to the contractor is all that has to be considered – traditional single stage competitive tendering provides a good basis for monitoring this scenario. However, although money (the tendered price) provides a common basis of comparison of the extent of the resources to be used by the contractor, the client may still have an unsatisfactory situation. If the tendering procedure and contractual arrangements are badly drawn up, the contractor may be using his resources inefficiently. This is dealt with in much greater detail later. To give a very simple example here: if a client has two similar projects, he will not necessarily get the lowest final price by setting up competition between one set of contractors for the first project and another set for the second and taking the lowest in each case. If the two projects are linked or combined in one, the price of the larger single contract may be less than the two let separately, because of the savings in building resources.

The first priority in tendering is to ensure the most economic use of the building resources, bearing in mind the particular needs of the client, and then to ensure that the price paid for those resources is as low as is reasonably possible.

The economical use of contractors' resources is becoming a much more difficult problem to assess today, as compared to when the building industry was mainly craft-based, and tendering reflected little more than the contractors' profit and their management skill in organising the output of their craftsmen to complete the job. In such a situation there was little opportunity for contractors to vary the deployment of their resources. With limited plant and machinery available and with production knowledge limited to the best way to organise gangs of craftsmen to carry out the work, it was possible for the contractor to be given the quantities and for the lowest tender then to represent the lowest production resources.

Today, there is still much work that comes into this category, but there are many elements in a modern building where different considerations apply. Even with the traditional elements of labour, materials, plant and capital, there are additional factors to be evaluated.

Labour

If labour is plentiful, a labour intensive form of construction will be economic, but, if labour becomes short, generally or in specific trades, difficulties may arise leading to inflated prices. It is also worth considering whether the emphasis can be shifted from site based, to factory based labour – a cold, wet and windy site in the middle of January is not the best environment to encourage good productivity from the workforce.

Materials

Not only should the use of new materials be considered, but also availability, bulk purchasing, special relationships between contractors and suppliers and special discounts. Life cycle costing studies have shown that increased investment in quality in original construction, can reduce maintenance and replacement costs during the life of a building, leading to better economics overall.

Plant

A balance has to be struck between the use of larger factory made components, needing extensive cranage facilities with hard access and a consequent saving in site labour, and the use of small units, with minimum site plant requirements. Continuity, or the lack of it, in the use of expensive items of plant, has a significant effect on the economics of production.

Capital

Cash flow is vital to most contracting organisations – keeping loan charges to a minimum can significantly affect contractors' tenders. Keeping retention percentages to a minimum, and the prompt reimbursement of fluctuations in the prices of labour and materials are relevant here. However, a balance has to be struck between the maintenance of a reasonable retention fund and easing the contractor's cash flow.

There is, of course, another element that is vital in the quest for the economic use of resources. The most careful planning of the use and co-ordination of labour, material and plant in the design stages of a project, will be of no avail, if the contractor does not have the **management expertise** to bring the master plan to reality in practical terms. This is particularly important in the planning and co-ordination of specialist sub-contractors' work. It is this aspect of project planning that has led to the extensive use of pre-tender interviews with prospective contractors – and even to partnering, which will be referred to in greater detail below.

Contractor's contribution to design and contract programme

There are occasions today when the contractor has a contribution to make which affects the design, speeds up construction, uses fewer production resources and, thereafter, produces a more economical job. The importance of this contribution should not be overstated, as a substantial majority of building works are, and will continue to be, small in value, using fairly traditional methods. However, it is a factor not to be ignored and it is interesting that, where sub-contractors are concerned, there is almost an undue readiness to consider the sub-contractor's contribution, whilst the main contractor's is often accepted with reluctance. This dichotomy is partly due to the fact that sub-contractors tend to represent specialist activities entailing considerable design responsibility, whilst main contractors frequently show reluctance to accept any such responsibility.

Engineering works differ somewhat from building works, in that the responsibility for detailed design is more frequently left in the hands of the contractor.

Cost planning techniques assist architects in designing economically. However, it is not always possible for the quantity surveyor or other members of the design team to know the best way in which the contractor's resources can be used to improve productivity in terms of cost and speed, particularly where a proprietary system is involved. For example, where a complete structure is fully designed before the contractor is chosen, the use of a proprietary system might well be excluded, despite its obvious economy, because of the subsequent amendments that might have to be made to the design, in terms of detailed fixings and finishes. It therefore becomes necessary to consider, at the pre-tender stage, whether a contractor has a contribution to make or not, and if he has, when this should be introduced. This matter is discussed in detail in Chapter 4.

Production cost savings

The economics of the construction industry's production methods are clearly a most relevant factor in the cost of building and obtaining value for money for the industry's clients. In terms of a particular project, this comes down to the economic use of the production resources held by a particular contractor. Production resources are dynamic, not static; production savings are continuously being made in industry, and the building industry is no exception.

Construction has special problems in relation to increased productivity where design is separate from production. Most industrial firms have a great incentive to make production savings as the firms themselves benefit. In the construction industry this incentive is frequently limited by the design and is, therefore, confined to more productive ways of carrying out the design. It is difficult to promote savings where the design has to be altered to make such savings. Responsibility still remains with the designer, who is naturally reluctant to make amendments which may put him at greater risk. Unless special contractual arrangements are devised, the contractor will not have the incentive to look for productivity savings where they would affect the design.

Where the opportunity of recognising production cost savings is considered to be an essential ingredient of a procurement programme, special arrangements have to be made, including creating the incentive for the contractor to do so. This is often done in conjunction with continuity contracts which are discussed below. Although this may seem to be a daunting task, it is only an extension of a contractor's normal bonus system, so contractors are very familiar with the principles.

Continuity

Continuity of activity is perhaps one of the most important ways in which production and management resources can be used economically and, provided the tendering procedures are right, the client can benefit. Building has a unique problem in relation to the provision of continuity, in that each building is on a different site, and frequently user and client requirements will vary as well. Nevertheless, a great many buildings are for clients who will want similar facilities on similar sites and, furthermore, at least some of the detail in dissimilar buildings will be the same.

Everyone takes longer to carry out a task the first time. Big savings arise the second time, rather smaller ones the third and so on until, after carrying out the activity many times, virtually no further savings can be

achieved. A succession of one-off situations, in management and production, is bound to be uneconomical, and this is one of the biggest
single areas in construction where better use of production resources
can be obtained with careful planning.

This factor has a very substantial effect on tendering procedures for, if
continuity from a particular contractor is to be achieved, then it must be
on a basis other than normal competition. This may make it necessary to
plan the contractual arrangements for an initial project with a view to
further projects being carried out by that contractor. Obviously,
productivity savings arising from such continuity must be translated into
a corresponding benefit to the client. Such methods as continuation
contracts, serial contracting and term contracts are methods of doing
this and are dealt with in Chapter 13.

Risk and accountability

Risk is also an important factor to consider in tendering procedure and
contractual arrangements. It is sometimes thought that provided the
contractor takes all the risk, accountability is satisfied. Frequently,
however, this satisfaction is delusive, the true situation being hidden.

Risk is defined as the possible loss, which has to be stood by someone,
resulting from the difference between what was anticipated and what
finally happened. It is important to realise that a building owner takes on
a considerable risk merely by the act of commissioning a building at all.
The contractor never takes the total risk. The client, for example, takes
the risk in relation to such matters as what the authorities will permit
him to build on the site. The design team take the risk for the work they
do, and so on. The question to be decided in the tendering procedure and
contractual arrangement is how much risk it is appropriate for the
contractor to take.

It does not always pay the building owner to ask the contractor to take
the risk. Insurance companies, whose very purpose is to take the risk for
you, charge a premium. There is no doubt that in every contractor's
tender there is a hidden premium charging the client for the risk he has
been asked to take, and in some cases, where the risk is very high, it will
not be worthwhile for the client to pass the risk to the contractor, just as
in some cases it is not worthwhile insuring against the risk.

It is significant that government does not pay premiums to insurance
companies to take the risk of destruction of government buildings by
fire. This is good practice in terms of accountability for public money.
Such are the resources of government, there is no need to pay others to
take the risk. On the other hand, for small organisations, the destruction

of a substantial asset can be disastrous and they must pay someone else to cover the risk for them.

In building, risk is much affected by the state of the client's own resources. It would be wrong to push the analogy with insurance too far. It is often difficult to distinguish the risk that is appropriate for the contractor to take, as an inherent part of his production activities, from the risk inherent in the construction itself. It is, of course, entirely appropriate that the contractor should take the risk for activities which are entirely under his control. It is appropriate, too, for the contractor to take the risk in operations, which are basically under the control of the building owner or his designers, where the building owner, because of his own circumstances, wishes the contractor to include in his tender a premium for taking that risk.

To illustrate this point, take on the one hand the conversion of an existing building where considerable risk is involved owing to the nature of the work. If the risk is too great, it may be impossible to get the contractor to tender at all. If it is moderate, then it would probably pay a client with substantial financial resources to take the risk himself and let the contract on a cost reimbursement basis. On the other hand, an individual of limited means, carrying out a conversion of his own house, may feel it is essential to get the contractor to take the risk and therefore get a fixed price.

A very common risk area today is that of fluctuations in prices of labour and materials, with the question as to which party should bear the risk involved.

Risk is one of the fundamental factors to consider in selecting suitable tendering procedures and contractual arrangements. It is a factor which depends on the client's circumstances as well as on the nature of the risk.

Accountability

An early reference to accountability was made in the above section considering risk, and the two subjects are very closely linked. However, accountability also has much wider implications and a full discussion is set out in Chapter 3.

Current trends

Although the general principles described above still apply, in recent years a number of new ideas have been introduced. These new ideas have included attempts to introduce new approaches to the procurement process, new techniques to improve the service to clients, proposals to improve relationships and performance and attempts to speed up the process – sometimes a combination of these factors is claimed. Among these new ideas, are the following:

- partnering
- value management
- extended arm contracting
- the New Engineering Contract
- the Private Finance Initiative
- the Housing Grants, Construction and Regeneration Act 1996.

Partnering

Partnering may be considered as the latest in a series of attempts to improve performance in the construction industry. By encompassing the advantages of continuity, production cost savings, parallel working and non-adversarial relationships, it is claimed that significant improvements in performance can be obtained.

Sir Michael Latham highlighted the concept in his report *Constructing the Team* in 1994, and the Centre for Strategic Studies for Construction, via the Reading Construction Forum, continued the theme in its 1995 publication *Trusting the team: the best practice guide to partnering in construction*. In 1997, the European Construction Institute published a document entitled *Partnering in the public sector: a toolkit for the implementation of post award, project specific, partnering on construction projects*.

There is therefore no shortage of published material on this subject and the reader may find it instructive to peruse these publications to gain an understanding of the different authors' perceptions of, and aspirations for, the arrangement.

The quantity and variety of published material available perhaps gives a hint to the nature of what is ultimately a flexible and very 'party specific' procurement system. Any attempt, therefore, to distil the essence of partnering is bound to miss some of the finer points incorporated in any specific arrangement.

Partnering has been described as a management concept, based on teamwork, used by two or more parties, to achieve specific business objectives, by maximising the effectiveness of each participant's resources. The concept is based on at least three key premises:

- mutual objectives – improved performance and reduced cost
- the active search for continuous measurable improvement
- a common approach to problem resolution.

Other contributors have extended these principles to encompass 'seven pillars of partnering', namely, strategy, membership, equity, integration, benchmarks, project processes and feedback. No doubt, the number of principles depends on the particular circumstances and the detailed agreement between the parties. These and other considerations are examined in greater detail in Chapter 14.

Value management

Value management is usually understood as the extension of the principles of value engineering to cover a project from inception to completion. Value engineering is a process by which a product, or process, is refined in terms of performance and cost, so as to produce optimum value for money. The process involves the systematic identification of the function of a product or process in the context of its use, to establish the lowest possible cost while still maintaining the required function or performance. It is a process of continual reassessment and refinement, as the balance of priority between quality and cost can change as the project develops or when variations are incorporated.

Value engineering can be applied to any product or process at any time – although the benefits are most significant when the process is applied early in a project and over as wide a range of activities as possible. The logical conclusion of this concept is to apply the technique from inception of the project – it is then termed *value management*, and can affect the whole design and realisation of the scheme, not just the production stage.

The process of value management integrates scientific analysis with subjective ideas, and therefore combines the ability to examine a process methodically, while applying a natural logic and subjective judgement based on practical experience, as much as theoretical practice.

The process, although sounding complex, involves a simple discipline which is adapted to the particular facet of the project being considered. It can encourage the use of creative and abstract ideas to produce radical changes in design, it can develop those ideas by way of a system of logical and progressive suggestions that arrive at the objective of eliminating unnecessary cost.

Although cost reduction is the main objective of the process, the elements of time and quality are not ignored. Essentially, this is the aspect which separates value management from other cost control systems. It provides a balance which ensures that a value study does not end up purely as a cost cutting exercise. Indeed the emphasis is on identifying the significance of the cost, quality and time aspects of a product or process and setting the appropriate priority to meet the client's objectives. By sensitivity testing (evaluating the 'what if' scenario, to assess the significance of changes in the equation) the client can be informed of the cost effect of marginal changes in quality and/or time and can make decisions accordingly.

The process is labour intensive, and therefore expensive to operate, but, on major projects, significant advantages can be obtained and the system can be said to at least pay for itself, if not provide a positive bonus. While this provides a comfort factor for the client, who pays the bills, it is incumbent on the client's advisers not to embark on such a process without proper consideration of the real need, or benefit being realistically attainable.

Extended arm contracting

Another process, by which the advantages of continuity can be harnessed, is extended arm contracting – a phrase coined to distinguish a slight variation in working relationships between the parties, usually in a management contracting or construction management environment, sometimes including an element of design. In this process, clients with continuous or considerable programmes of work, seek to optimise the services of tried and trusted contractors – building on previous experience. By extending the working relationship with a contractor who has already demonstrated compatibility, the client reduces the learning curve and gains operational and economic advantage.

This is a particularly useful system where the working environment imposes severe constraints on the participants – as in motorway improvements and railway works, where the continued use of the existing facilities is of paramount importance. In this scenario, a contractor unused to the environment incurs huge overhead costs in 'getting up to speed' – such costs have to be recouped, which means that somewhere, someone is paying the price. This is not in the best interests of the economic use of resources in the national context.

Under such arrangements, the client often employs a quantity surveyor to oversee the process and monitor the contractor's performance – somewhat akin to an auditing process. As with many of the 'current trends' which rely on mutual trust and professional integrity, it is important to make sure that the system does not become too cosy, or that the participants lose their hunger in the pursuit of excellence on behalf of the client.

The New Engineering Contract

Although this book and the other AQUA books are generally based on the use of the JCT family of contracts, it is appropriate to comment on some aspects of the New Engineering Contract while considering current trends.

It is claimed that the New Engineering Contract embodies all of the criteria for improving contractual relationships identified by Sir Michael Latham, in his report *Constructing the Team*. The concepts behind this family of documents can be summarised as follows:

- The contract is based on a spirit of co-operation, rather than adversarialism. Because of this, there are those who consider that it is not legally watertight and that an unscrupulous contractor might 'drive a coach and horses through it'. This is a risk, if the parties are not committed to the basic concept of the form.
- The contract is drafted in simplified language, to encourage its use in parts of the world that do not understand the niceties of English legalese.
- Under the concept of co-operation, rather than adversarialism, the form is kept comparatively short and much of the administrative detail (which is embedded in the JCT contracts) is contained in Guidance Notes, to become custom and practice, rather than a contractual requirement.
- No priority between contract documents is identified.
- The basic form, through a system of alternative wordings, can be used in a number of different scenarios – with BQs, with specification and drawings or with activity schedules, and as a fixed price or cost reimbursement contract, in traditional form, or as a target cost or management contract. The use of a single form avoids the problem of identifying different clauses in different forms – the common clauses are either in use or deleted.
- Users have to get used to the absence of cross-references between clauses.

The Private Finance Initiative (PFI)

In a climate of reducing public expenditure, government has encouraged the private sector to extend its philosophy of design and build into the public sector. The overall logic of this concept has been much debated, as it can delay projects, and possibly increase costs (as opposed to government expenditure) rather than reduce them. However, the system is firmly rooted in the industry and is unlikely to disappear in the foreseeable future.

The concept is that the private sector, usually in the form of an *ad hoc* consortium, takes on the responsibility of the design and building of a capital project, which is then sold, or perhaps leased, to the public sector when complete. Thus the risk inherent in the design and build process is passed on to the contractor or consortium. The main difference between this system and normal design and build, is that under the PFI, the responsibility for funding the project is also passed to the contractor.

This leads to the apparent anomaly in terms of overall cost – for, assuming that no organisation can borrow money any more cheaply than the government, the cost of the funding, and the risk involved, must attract a premium from the contractor, which is not present in traditional procurement. The assumption must be that the system affords other economies, which outweigh this premium – however, such is the nature of PFI projects, and such is their programming, that the actual evaluation of the comparative overall cost can never be demonstrated. The main attraction to government lies in the delay of the capital expenditure, or the transfer of the cost from capital expenditure to revenue expenditure, especially in the case of DBFO (design, build, finance and operate) schemes.

Experience of PFI schemes has not been extensive, because of the small number of schemes that have come to fruition. Much more in evidence is the frustration of contractors and *ad hoc* consortia that have made tremendous investment in the system and have had great difficulty in getting approval for schemes to proceed. This delay is, perhaps, part of the overall planning and control of government expenditure.

Parties entering into such arrangements must be aware of the commitment involved, the risks inherent in the system and the funding or cash flow implications. In other respects, the system requires the usual application of the principles of good practice advocated in this and other AQUA books.

The Housing Grants, Construction and Regeneration Act 1996

The Construction Act, as this unwieldy statute has become known, is another manifestation of the effects of Sir Michael Latham's report *Constructing the Team*. It is an attempt to improve working relationship under construction contracts.

There can be few involved in the construction industry, including designers and other consultants, who are not encompassed by the Construction Act. How much they will be affected by it, remains to be seen.

There are, however, some significant exemptions to the Act, as follows

- contracts with a residential occupier, actual or intended
- contracts not in writing.

The principal provisions of the Act are two-fold. Firstly, to introduce adjudication, and secondly, to introduce payment provisions into all 'construction' contracts. The implications of these two provisions are almost universal as far as those involved in the construction industry are concerned, except for the two exemptions identified above. It is important to note that contracts must make their own provision for these matters, sufficient to comply with the Act, otherwise the provisions in the Act, or in relation to adjudication, the provisions of the 'scheme for construction projects' come into force.

The impact of these provisions has yet to be felt to the full – certainly adjudication will take its place in the normal sequence of events in the resolution of disputes. However, the extent to which they will affect consultancy appointments is less clear – one aspect which is certainly exercising the minds of consultants and their insurers is the effect of the adjudication provisions on consultants' professional indemnity insurance policies.

Certainly, the introduction of the Act has changed the environment in which the industry works, and all those involved in it must take cognisance of the requirements laid down.

The detailed effects of the Act on the pre and post contract activities of members of the construction team are matters to be considered in the context of AQUA books *Pre-Contract Practice for the Building Team* and *Contract Administration for the Building Team*.

Conclusion

For convenience, set out below are the five fundamental factors to be considered in the selection of a contractor and the type of contract to be used. To some extent they overlap and they certainly interrelate.

(1)   the economic use of building resources
(2)   the assessment of the contractor's contribution in relation to the design and speed of construction
(3)   the incentives to make production cost savings and their control
(4)   continuity of work in all aspects
(5)   risk and the assessment of who should take it.

In addition it is worth noting the distinction between the selection of the contractor and the determination of the contract price. The traditional way of selecting a contractor, that of full drawings coupled with a bill of quantities, describing and quantifying the work to be done, results in the selection of the contractor simultaneously with his price being known. In this situation, the drawings and bills of quantities are tendering documents and also rank as contract documents. Selection and determination of price occur together.

In many other situations this is not possible. What in fact happens is that the selection of the contractor is made, but only the basis on which the contract price is to be obtained is agreed. The final determination of that price may take place many months later. Indeed, it does not have to be determined before the contract is made. If this point is appreciated, some of the more unusual methods of tendering will become easier to understand.

Having identified the five fundamental factors, and the principle of splitting the selection of the contractor from the determination of the contract price, it should be easier to understand the concepts behind the latest trends in procurement. As government policies and commercial pressures change, the ingenuity of the members of the construction team seems to know no bounds, and subtle, and sometimes significant, changes in the procurement process emerge. It is anticipated that, armed with all the techniques discussed in this chapter, the construction team can be more efficient in identifying the most appropriate method of procurement in any given set of circumstances, and thereby provide the best service to the client.

Finally, it should be emphasised that although all these factors should be considered for every project or group of projects, it is recognised that in a great many small projects, particularly those which are completely one-off and not part of a larger general programme, some of them will hardly apply. Tendering procedure must be considered in the context of the national scene. Most building contractors are relatively small and the majority of builders would be unable and unwilling to tackle a job of any size. Many small contractors, while making a first–class job of constructing a building to an architect's design, have neither the wish nor the ability to make the kind of contribution which a few national or specialist firms might make in some of the ways identified in this and subsequent chapters.

'... a matter on which the employer should have an opinion ...'

Chapter 3
Accountability

Chapter 2 considered five main factors in selecting a contractor and formulating arrangements for a satisfactory contract. Accountability is a further factor which can have a profound effect on these matters and such is its special nature that it deserves a chapter on its own.

Accountability arises whenever one party (the agent) carries out an activity on behalf of another (the principal or employer). The agent must account to the principal for the actions he takes. The extent of that account (or level of accountability) must depend upon the principal's original brief and the degree of authority and responsibility delegated to the agent. In building procurement, that authority and responsibility lies within a framework of established principles, with appropriate priority being given to cost, quality and time.

Background

Historically, the emphasis has been on accountability within the public sector, although its relevance should be no less important within the private sector. The difference is purely one of reporting lines. In the private sector, the relationship between agent and employer is direct: any matters can be settled between the two parties. In the public sector, the responsibility of the agent to the employer (the public) is through an intermediary, usually an elected body, who will normally be advised by professional staff and who will ultimately be monitored by an independent audit body. There is normally no direct relationship between the agent and the public as employer.

The accountability of the professional adviser is an important aspect of the tendering procedure in construction, because it is seldom possible for the building owner to see what he is buying before it is built. Even when this may be possible to some extent, for example in the case of a standard building, each will be on a different site and thus becomes unique.

Accountability differs from the other factors concerned with the tendering process, all of which are technical aspects, requiring professional advice to determine their effect. Accountability is a matter outside the strict building process, since it applies wherever the employer's agent is acting on his behalf in respect of purchasing, e.g. buildings, vehicles or stationery, etc. Clearly, therefore, it is a matter on which the employer should have an opinion and which should be considered to be part of his brief. Professional advisers, as agents, are responsible and accountable to the employer for both their actions and performance.

The modern concept of public accountability

The traditional concept of public accountability dates back to the Middle Ages. At that time production was craft orientated and the sophisticated methods of modern production and construction were unknown. Problems related to production where insignificant. If the State voted to spend money on ships for its navy, the problem uppermost in its mind was to ensure that the money voted for the ships was actually paid to the shipbuilder, rather than finding its way into somebody else's pocket. The method used in those days for the placing of orders was probably simple bargaining, implicit in which was the objective of value for money. The simplicity of the whole operation served to highlight the only points where the matter could go seriously wrong; these related entirely to graft and corruption.

As modern methods of production and construction have developed, it has become more important to be able to show that value for money has been obtained. While the standard of honesty of public officials is generally high, the constraints imposed upon them, to ensure that there is no corruption, tend to work against the concept of value for money. Indeed, it is now possible for all the strictest canons of public accountability to be adhered to, and yet extremely poor value for money to be obtained, due to the waste of resources inherent in present public accountability procedures.

The main reason for this is that, with complex technological production methods, the inefficient use, and therefore waste, of resources easily arises. The resources that can be wasted in this way, can well exceed those which, in former times, found their way into corrupt pockets. To avoid such waste, positive steps must be taken to ensure the optimum use of contemporary procurement and production techniques.

Contract documentation

The amounts of money involved in construction work are usually large and this means that money can be more easily misappropriated – thus there is more opportunity for corruption than in many other areas of activity. Contract documentation for construction work is therefore very important – it is vital to the concept of obtaining value for money and may be complex and technical, varying with the system of procurement adopted (see also 'Dispensing with competition' below).

Proper price

The client should have an assurance that he is getting value for money or paying the proper price for the project even though it may not be possible to determine its total extent, or final price, at the time of the signing of the contract.

Where tenders are sought on a competitive basis, with full contract documentation, the lowest price is normally accepted as being the proper price for the work, since all tenders are related to a common base. Where full documentation is not provided (e.g. in plan and specification or design and build contracts), although the basis of tendering is still common, it can be more difficult to convince a client that he is paying the proper price, because the precise specification may not be clear at tender stage.

Dispensing with competition

There may well be times when the use of competitive tendering is not possible, or will not necessarily produce value for money. There may be a number of reasons why the contractor should be selected without competition, and they may be complex and intangible. In many cases, it may be difficult to prove that there will be a monetary saving without the use of competitive tendering and the agent will have to convince the principal by justifying his professional judgement.

Where a contractor is selected on a basis other than equal competition, the contract sum or pricing mechanism must be negotiated and the matter becomes even more complex, especially if full documentation is not available. Not only does the selection of the contractor have to be justified, the concept of the proper price, or value for money, has to be demonstrated.

Here again the documentation is vital – depending on the time and information available, a system must be set up that will not only allow a contract sum to be calculated, but will also establish a pricing structure for the assessment of interim payments and variations, where appropriate.

By their very nature, negotiated contracts often involve uncertainty, and the agent has a special responsibility to the principal to explain or justify the professional judgement that has led to the method of procurement recommended.

In the public sector, whether in central government, local government or other government controlled organisations, it can be difficult to make a convincing case when accounting for the decisions thus made. Generally speaking government, in its widest sense, will always seek to obtain tenders on the basis of competition, as there is always the danger, however slight, that decisions leading to the selection of a suitable contractor by any other method may be subjective, and possibly corrupt (or appear to be so).

Inflation

A further factor related to accountability arises from inflation. Particularly in times of high inflation, the argument has sometimes been advanced that the tendering procedure should be arranged so as to bring forward the commencement of building in order that an earlier price level may be obtained (to beat inflation). This may be a real saving to a private client, depending on his circumstances and the use to which his money is put in the meantime. To government, however, whose concern is the percentage of resources devoted to building, there is no saving at all, as inflation affects both sides of the balance sheet. However, this fact needs to be weighed against other factors obtaining at the time; public finance accounting procedures and the way in which central government organises its annual financial allocations may well place a premium on early financial year starts within the public sector.

Value for money

Value for money, as a concept, may be defined as achieving the optimum use of resources – resources comprising money, manpower, time and materials, each of which may be regarded as equally important. Until comparatively recently money was often regarded as the most important resource in the public sector, the cheapest method of construction being the one most usually selected. With money becoming more 'expensive' the need to obtain more for a given amount became imperative, and this has tended to emphasise the equal importance of constituent resources. For example, if a contractor is overpaid £1000 on a contract through an accounting error, there is a monetary loss of £1000; if, through design errors which do not manifest themselves during the contract, there is a loss of £1000 of building labour and materials, then again, there is a monetary loss of £1000. However, in the first case the money represents only financial resources which are no longer available for beneficial use, but, in the second case the money represents not only a financial loss, but also a loss of physical resources which have been used, wasted and lost.

Within both the public and the private sectors, the overpayment loss will be obvious; the loss of manpower and material resources may not. In either case, the professional advisers will be accountable to the building owner, to whom they are responsible, but whereas the former can be expected to be easily identified and recovered, the latter is considerably more difficult to detect and may never be recovered.

Government is concerned with the economic use of resources by the country as a whole. More resources used in building each project, means fewer projects. While government is concerned with the monetary waste inherent in overpayment, this is very much the lesser of the two evils. Unfortunately, traditional systems only bring the overpayment to the fore, while the loss of resources remains obscure. Politically, too, overpayment is an easy matter to grasp and is highlighted in political debates as an example of government ineptitude, while the loss of resources is difficult to identify and for this reason is more often forgotten or ignored.

The definition of value for money has broadened over recent years. As money has become progressively dearer, so the concept has widened beyond that of obtaining best use of resources on site (usually by acceptance of the lowest tender) to embrace the whole project from initial procurement, through design and construction, to maintenance and cost in use, in order that the best use/returns can accrue over the lifetime of the building. With this change in definition, terms such as 'life cycle costing' and 'value engineering' have come into vogue, with value defined as 'the optimum balance of time, quality and cost'.

Numbers tendering

Another important aspect of accountability, is that of the number of firms invited to tender for a contract. In *Pre-Contract Practice* it is advocated that, where competition is required, selective tendering should be used rather than open tendering. It is also suggested that a limit should be placed on the number of contractors tendering in accordance with the recommendations of the Code of Procedure for Single Stage Selective Tendering.

The arguments here are simple. If thirty contractors rather than five are persuaded to tender, the client might get a lower price on a particular job. On the other hand, if thirty contractors have to tender on all contracts, there would be an enormous waste of estimating and managerial resources. Each successful contract would have to bear the burden of the tendering expense for an average of twenty-nine unsuccessful ones. It is salutary to consider that even with a limited list containing only five contractors, the successful tenderer has, on average, to bear the cost of four unsuccessful ones. Inevitably this increases the tenderer's overheads which, in turn, are reflected in subsequent tenders – the client pays eventually.

Conclusion

The problem, particularly with public building, is to identify whether or not there will be a saving in resources in relation to the way in which the contractor is selected. This is a complex and difficult task, but in the long run the advantages to be obtained from many of the situations referred to later in this book will not be forthcoming unless the attempt is made.

Accountability is an extremely important matter in tendering procedures and contractual arrangements. It is essential that it is looked at in terms of all the resources used and not simply in terms of the price paid. It is essential for the client's professional advisers to identify the effect that the procedure for the selection of the contractor may have on the use of all building resources, and to ignore, or certainly question, the historic traditions relating to public accountability.

'... early selection ...'

Chapter 4
Early Selection: Two-Stage Tendering

Chapter 2 indicated various factors which should be taken into consideration in selecting a contractor. Some of these relate to the need to select a contractor early, prior to completion of the working drawings and bills of quantities. This chapter considers the occasions when early selection may be advantageous and how it should be achieved.

When to select early

Although early selection has been defined as selection any time before complete documentation is ready, in practice it has advantages only when done at a very much earlier stage. Selection after completion of documentation in the normal manner is preferable, unless those advocating early selection can truly justify its use. Basically there are two main reasons for doing so:

- to allow the contractor's participation in the design and programming process
- to allow the contract to start, and therefore complete, sooner.

Early selection to allow the contractor's participation in the design process

The contractor's contribution to the normal pre-contract process can improve the quality of the end product and might be recommended for a number of reasons:

(1) *Technical contribution to design*
Firstly it is essential to make an objective assessment of the value of the contractor's contribution: whether it will really amount to something positive which cannot otherwise be achieved, or whether it is more apparent than real. For example, if it were

intended to build several identical industrial units, the use of a proprietary cladding system might be justified. In such a case, it would be unwise to process the design through to final production without having carried out a preliminary cost appraisal of the proposed system and ensured that the system can accommodate the architect's requirements. It may be that the architect's details need to be modified to allow the system to be used economically. Then an objective assessment must be made to determine whether such a system can produce a more economical or acceptable solution than the use of a more routine non-specialist cladding material.

(2) *Management skills and buildability*
      The contractor's input, at the design stage, in terms of management and practical skills, can often be of benefit in avoiding construction difficulties on site, and assuring buildability.

(3) *Shortage of materials or time*
      Once a contractor is appointed he is able to use his expertise in procurement to place orders and obtain materials which may be in short supply, or for which there are long delivery times. This helps to minimise the risk of delay during construction. Additionally, being involved in the design process he can become more aware of the designer's intention and offer alternative construction solutions which save time on site.

(4) *Design alterations to suit specialist plant*
      Prior knowledge of the contractor's expertise with particular plant may make it possible to design in such a manner as to take advantage of its use, and so speed construction; for example, in the use of 'slipform' techniques in the construction of the concrete core of a multi-storey office block.

Figure 3 shows a diagrammatic representation of the different sequence of events when two-stage tendering is used to allow the contractor to contribute to the design process. The vertical divisions of the matrix are not intended to indicate any particular time interval – they are provided purely to assist visual comparison of the alternatives. Neither is the length of the activity bars necessarily significant, except in allowing a comparison to be drawn between the two systems. In practice, the duration of any activity must be agreed between the participants, so as to achieve the employer's expectations. Figure 3 indicates that no time need be lost in securing a contractor's contribution to the design – indeed, as indicated below, time may be saved.

Single-stage competitive tendering

Design			▬▬▬▬																		
BQ production				▬▬																	
Tender					▬▬																
Select contractor						▪															
Construction							▬▬▬▬▬▬								▬	▬	▬				

Two-stage tendering

Design			▬▬		▬▬																
Tender documents (Stage 1)			▬																		
Tender				▬▬																	
Select contractor					▪																
Negotiation (Stage 2)						▬▬▬															
Construction							▬▬▬▬▬▬							▬	▬	▬	▬				

Figure 3. Sequence of events in single and two-stage tendering.

Early selection to achieve earlier commencement and completion

The employer may wish to start work on the site, before production information is fully available, to achieve:

(1) *Early commencement*

A subsidy or grant may be directly linked to a statutory date for starting work on site. In the context of annual accounting, projects are sometimes only sanctioned if a start can be made before the end of the financial year. In these cases an evaluation has to be made between the possible extra costs involved in a start before design is complete, and the benefit of a grant. Early selection and parallel working (see below) may be preferable to an inefficient competition based on inadequate information.

(2) *Early completion*
The sooner the start, the sooner the completion, provided that design work and construction are programmed and integrated properly. A short overall programme of design and construction also assists in reducing the employers' interest payments on funding, and site costs. In this case, the contractor will not necessarily be participating in the manner described above, but could be working in parallel with the designers. For example, the contractor might be building the foundations while the final details of the structure are being designed, or erecting the structure while the details of the finishings are being finalised.

Such a system has its dangers; hence the need for detailed programming. It may be even more expensive in terms of construction, and some discrepancies between work in hand and work still to be detailed may perhaps arise. However, the income derived from the building investment may well make it worthwhile to save even a few weeks in the overall programme. Income gained could be used to help fund later phases of the scheme, which could perhaps follow a more traditional procurement route. Once again, an objective financial assessment must be made of the possible extra cost resulting from increased speed and the benefit of the increased revenue.

Figure 3 shows two-stage tendering used to allow the contractor to contribute to the design process. If the system is used to achieve an earlier start on site (rather than to allow the contractor to contribute to the design), the duration of the various activities will change. Depending on the specific circumstances of any project, the amount of time needed for the initial scheme design and the stage 1 tendering process may be reduced, thus allowing an early negotiation and an early start on site. If a cost reimbursement form of contract is acceptable to the employer, a very rapid start on site can be achieved.

How to select: two-stage tendering

It is possible, in any of the above cases, to select a contractor and carry out a single negotiation on the basis of cost alone. Some of the occasions when this might be appropriate are described later in Chapter 6, Competition and Negotiation. Whilst this may satisfy the requirements for speed and quality, it can result in increased risk to the employer with respect to the final negotiated price.

However, the process of two-stage tendering usually seeks to balance the cost–time–quality triangle, and can be used where early selection is required, but where there is no need to negotiate with a single contractor, thus avoiding some of the risk inherent in negotiating with one specific contractor.

The term two-stage tendering essentially describes the procedure where a contractor is selected in one operation and the contract sum is agreed in a second operation.

The first stage, the selection of the contractor, is conducted on a competitive basis in the pricing of brief, but precise, documents related to a preliminary design. This provides the basis of pricing for use in subsequent negotiation, but not a contract sum.

In some instances, the first stage may be preceded by a preliminary stage which involves interviewing selected contractors to establish the resources they have and the contribution they might make. It is undesirable that this preliminary stage should become the end of the matter. It can only be a subjective assessment leading to the first stage tender.

The NJCC Code of Procedure for Two-Stage Selective Tendering states that:

In order to minimise the cost of tendering overheads and to facilitate the selection of a contractor at an early date, documentation for the first stage should be kept to a minimum consistent with the need to:

- provide a competitive basis for selection;
- establish the principles of layout and design;
- provide unambiguous pricing documents, related to preliminary design and specification information, in a form sufficiently flexible to be suitable for pricing the second stage tender. Provision should be made for adjustment of price fluctuations during the period between first and second stage tenders;
- define the obligations and conditions of, and state the programme for, the second stage procedures;
- establish the intended Conditions of Contract.

Stage one: documentation and tender

Three methods of preparation of the documentation for the initial selection are set out below. However, there are many others and it is up to the design team to choose the most appropriate, bearing in mind the employer's requirements and the particular circumstances. The three methods briefly described are:

(1)   The general estimating method
(2)   The approximate quantities method
(3)   The cost plan/proprietary design method.

(1) The general estimating method

This approach is most suitable for large contracts of a general nature, where no particular design contribution is called for from the contractor. The tendering documents for selection of the contractor should include:

(a)   A preliminaries section, giving particulars of all site management costs and head office overheads and profit, etc. This can normally be priced on a fixed price basis, provided the general scope of the project is known. It need only be adjusted later if the scope of the scheme is materially altered.

(b)   The contractor's allowance for head office overheads and profit in relation to:

- his own work
- his own sub-contractors' work
- nominated sub-contractors' work.

(c)   The basis of labour rates. This can either be in terms of a calculation of an all-in labour rate, or alternatively of a net labour rate, with some of the indirect labour costs calculated as a total sum to be allowed against a total calculation of the labour required for the job.

(d)   Some means whereby the productivity of labour can be assessed. This could be a breakdown of pricing for typical bill of quantities items, with notional quantities relating to work the contractor is likely to carry out himself.

(e)   The method of dealing with sub-contractors' work.

(f)   The method of dealing with materials, making allowances for waste in typical cases.

Refinements can be made to the above list and details can be requested in tendering documents, for items likely to have an important effect on the cost of the contract.

(2) The approximate quantities method

This method forms an alternative to the general estimating method and may be used in similar circumstances. However, whereas the general estimating method establishes labour constants and material prices, the approximate quantities approach establishes competitive prices for measured items.

The first stage documentation will comprise a bill of approximate quantities, including the specification for materials and items which are likely to occur in the final design. The quantities stated are approximate, in that they have been estimated from the early design details. All the sections of a normal bill of quantities would be included, e.g. preliminaries, preambles, measured works, etc.

After a contractor has been selected and the design completed, the approximate quantities can be confirmed and the contract price established. It is important that the initial approximate quantities are as near to the final 'designed' quantities as possible, so that the level of overheads, plant and profit originally included by the contractor does not have to be re-negotiated or adjusted.

(3) The cost plan/proprietary design method

This method is normally used when the contractor has a fundamental contribution to make to the design, through the use of a proprietary system. The object is to select a contractor whose proprietary system best meets the requirements of the architectural scheme, and to judge the offers received in terms of best value for money.

With this method it is important to have preliminary interviews with the contractors, to establish which have systems that can be adapted to suit the architect's sketch plan. There is no point in asking a contractor to tender if he has to alter his system uneconomically, or if the design has to be restricted by the limitations of his system. A number of interviews with possible contractors may be necessary to select a short-list of tenderers.

In preparing tendering documents, it is essential to identify the areas of work under the following six main headings:

(1) the works to be carried out in a traditional manner, such as external work, external services and possibly the foundations
(2) the work included in the proprietary system
(3) the nominated or approved sub-contract work
(4) the fixing or assembly of suppliers' work in the superstructures
(5) preliminaries and site management
(6) head office overheads and profit.

By the very nature of this work it is not possible to identify these areas accurately. Some proprietary systems take in more work than others. The tendering documents must provide for flexibility, so that the contractor who can economically incorporate a certain piece of the work in his proprietary system will be able to do so. However, the architect will quite rightly wish to have as much control as possible over those items which are not an essential part of the proprietary system. The following documentation is appropriate:

- Bills of quantities, or approximate quantities, for the external services and external works. The actual limit of measurement will depend on the circumstances, and each case must be judged on its merits. The external elements are normally unaffected by whichever proprietary system is adopted, and the architect is free to specify these as he chooses.
- Bills of quantities or approximate quantities for the elements unconnected with the proprietary system, e.g. foundations. Care is needed because, in some cases, these elements can be designed independently of the system, but this is not always so.
- A performance specification for the work covered by the system should be prepared. This must have built-in flexibility to allow for acceptable variations between each system. Wherever possible, provision must also be made for a breakdown of the price for the system content – this is for control purposes, once the contractor has been selected. If included, quantities must be on a functional basis, e.g. areas of floor related to superimposed load.
- The remaining work will largely be covered by sub-contractors, who will either be nominated or agreed and approved between the architect and the main contractor. This area of work is, for tendering purposes, best covered by giving details of the cost plan quantities, with a specification for each element. The contractor can then see what is required and if his proprietary system encroaches on this work. This is important because it allows the price for the area of building covered by the proprietary system to be seen in the context of the building as a whole.
- There will be certain items of builder's work in the superstructure, such as the fixing of doors and fittings, which can be covered either by a schedule of rates or alternatively by approximate all-in quantities.
- Site management costs and head office overheads and profit will be dealt with as fixed price items, as in the general estimating method.

Evaluation and selection of proprietary system

Whatever the format of the first stage, when an offer is received, it will not appear as a straightforward tender figure on which an immediate comparison can be made with the other tenders. An objective evaluation has to be made of the various tenders, weighting items in order to arrive at a common denominator. For example, a contractor who has decided to use two tower cranes instead of one will have included a greater amount for this item but may be saving labour.

Similarly, whilst one proprietary system may be more expensive than another, its effect on the remainder of the building may result in an overall saving. Conflict may arise between the objective assessment in financial terms and a subjective opinion on performance or aesthetic grounds. When this occurs, it is essential not to mix the assessments. Those matters which can reasonably be evaluated in terms of money should be assessed in that way. But overall, the cheapest tender may not provide the best solution. In such cases evaluation is difficult and, however objectively carried out, it will be bound to contain some element of subjective assessment.

In making the evaluation, it is worth pointing out to the employer the extent of the area of opinion as against straightforward objective assessment, and the possible effect this could have on the selection. Additionally, once the contractor is appointed, the employer has to pay the fixed amount of overheads and profit, irrespective of the amount of work to be undertaken. Therefore they carry as much, if not more, influence and weighting in the evaluation as the other rates and pricing in the first stage tender.

Stage two: determination of the contract price

Having evaluated the offers and advised on selection, it is necessary to proceed to the second stage, the determination of the contract price. This should be done as early as possible, but must be tied to the evolution of the design. Where a proprietary system is involved there could be a considerable amount of design refinement to tailor the system to the architectural requirements. During this period of evolvement, negotiations on the price within the context of the tender will be carried out, and finally a firm contract price will be agreed reflecting the finalised contract drawings and documents.

The second stage documentation format should closely follow that used for the first stage. The whole second stage document must be in a form to which cost control can continue to be applied during the construction stage. The second stage document will be priced on the basis of the first stage pricing data. If the programme is urgent, entailing parallel working, it may be necessary to agree the pricing on stage two progressively.

In meeting a tight timescale, often experienced in the second stage negotiations, care should be taken to avoid compromising the quality of the final stage two documents. Equally, care needs to be taken to avoid agreeing uneconomic prices – provisional allowances should be included, if full details are not available. Properly negotiated prices can then be agreed when the full information is available.

Advantages and disadvantages

In the appropriate circumstances, two-stage tendering has a valid part to play as an effective procurement procedure. However, there are both advantages and disadvantages to its use which must be made clear to the employer before it is adopted.

Advantages of two-stage tendering

- An overall tender price is usually known before work commences.
- There can be a saving of time over single-stage tendering methods due to the tender/design overlap prior to a site start.
- Control of the design and workmanship on site is maintained by the employer's consultants.
- The procedure still enables full tender competition of sub-contract works to be carried out as a parallel operation.
- Fewer disputes should arise as the contractor has been more involved with the design and document preparation, and has a better understanding of the requirements.

Disadvantages of two-stage tendering

- The contract sum may be more expensive than the single-stage method, because of the negotiated element and reduced level of competition.
- Delays in providing design information can result in project overruns and additional costs.
- Second stage tender negotiations could prove difficult and protracted. In such circumstances, a higher than market price may have to be accepted if delays are to be avoided while alternative tenders are obtained.

'... *someone from a different profession* ...'

Chapter 5
Tender Documentation
(Drawings, Specifications and Quantities)

In Chapter 1, in the section on contract conditions, a number of different contracts commonly in use were listed, and, later in the book, specific formats are looked at in depth.

One of the basic criteria for determining the type of contract best suited to the employer's needs is the preferred timetable, coupled with the information that is, or can be made, available for tendering within that timetable. In Chapter 8 there is a detailed tabulation of the various options available in this respect in relation to fixed price contracts, which probably still form the contractual basis for the majority of projects.

Normally, information to enable a contractor to construct a project is set out in drawings (usually taken also to include schedules) and a specification. When bills of quantities are used, the specification is usually included as preambles in the bills and the administrative details relating to the contract are set out in the preliminaries section of the bills. If bills are not used, the specification becomes an independent document and must include the administrative details. All are essential as part of proper tender and contract documentation.

Traditional arrangements

Good practice in the production of drawings, specifications and quantities, in the context of traditional working relationships, has been set out in detail in *Pre-Contract Practice* and there is no need to repeat that information here. What is required, however, is an appreciation of the significance of the various documents and their contents in non-traditional contractual arrangements.

Non-traditional arrangements

When the traditional mould is broken, the same information is still required to complete the project, but it may have to be provided at a different time, by a different party and in a different format – depending on when tenders are sought.

Non-traditional arrangements may include

- cost reimbursement contracts
- target cost contracts
- guaranteed maximum price contracts
- management contracts
- construction management contracts
- two-stage tendering
- design and build contracts
- turn-key contracts.

Set out below are those factors which must be taken into account when considering the use of drawings, specifications and bills of quantities in non-traditional arrangements.

Drawings

The purpose of the drawings is to illustrate the employer's requirements at each stage of the procurement process. Thus, they will serve to refine the employer's brief; they will allow the quantity surveyor to prepare cost estimates and eventually a cost plan; they will be submitted to obtain statutory consents, used to obtain tenders and, finally, they will be used to instruct the contractor what to build.

Preparation of the drawn information therefore passes through many stages and the point at which they are required for obtaining tenders, as contract documents or as working drawings may not always be predicted. This will depend upon the employer, his programme, the financial needs of the development, its size and the anticipated form of contract, all of which should be discussed at the earliest opportunity. Decisions taken during this process will dictate who prepares the drawings required at each stage.

Planning the course of the project from the day of its inception is essential if the process is to be free of claims and disputes. Proper planning of the process should ensure that the team is in control of the project and will be able to adapt the procedures should the need arise – even to the extent of changing the procurement path chosen.

Planning the production of drawings, be they architectural or engineering, must allow time for the necessary design processes, including analysis of the brief and consultations with appropriate authorities and specialists. Time must also be allowed for the preparation of cost estimates and bills of quantities, if they are to be part of the procurement system.

Drawings and schedules are fully discussed in *Pre-Contract Practice*. However, as a part of the documentation required for tendering purposes, their form and content related to the selected method of procurement may be briefly summarised as follows:

Table 1 Progressive refinement of documentation.

Tendering purpose	Description of drawings and other documents
Term contracts	Drawings may not be available or appropriate. For maintenance term contracts, a schedule of work items may suffice
Some cost reimbursement contracts	Time may preclude the production of drawings (e.g. emergency repairs) or they may not be available until investigatory work is completed (e.g. fire reinstatement)
Early selection of contractor by two-stage tendering Design and build contracts (employer's requirements) Some cost reimbursement contracts	Site and general plans, elevations and sections, illustrating the employer's requirements, supported by outline specification. (Scheme design, suitable for preliminary cost estimating and discussion with consultants or obtaining planning consent) If on CAD, layered for extension of contents and addition of information
Management contracts and contracts substantially dependent on specialist sub-contract design work 'Fast-track' arrangements Target cost contracts	The above drawings extended to form general arrangement drawings for use in obtaining statutory approvals, preparation of approximate bills of quantities and use on site; interfaced with services and structural engineer's schematic layouts and supported with a selection of typical and special details; suitable for cost planning and obtaining sub-contractors' and suppliers' preliminary quotations
Fixed price contracts based on single stage competitive tendering Negotiated contracts (second stage of a two-stage procedure) Works packages (under management contract) Contractor's proposals (under Design and Build)	General arrangement drawings and schedules fully co-ordinated with structural and services engineering drawings and annotated and cross referenced. Room layouts, large-scale part sections and elevations of special areas, all typical and special construction details; full specification; all suitable for the preparation of bills of quantities, for use in purchase negotiations and for construction on site

The essence of this summary is that, as the requirement for price certainty increases, or the time for construction approaches, the more complete must be the drawn information available.

It will be noticed that, depending on the procurement path chosen, the responsibility for the production of drawn information may change. Under traditional procedures, consultant designers are responsible for the information from start to finish. In some of the procedures outlined above, that responsibility may pass to the contractor. If that responsibility does pass to the contractor, it may be that the same designers are involved, but that they are employed by a different principal – this involves novation of the design contract, which is explained more fully in Chapter 12.

Specifications

In *Pre-Contract Practice*, there is a complete chapter on the subject of specifications and there is no need to repeat that information here. What follows is an explanation of the use of specifications in non-traditional arrangements.

Drawings and specification-based documentation

There are cases in which bills of quantities are not required and then the tender and contract documentation will comprise drawings and a specification. Subject to certain qualifications made later with regard to performance specifications, we do not recommend that a specification with drawings be the only tendering documents for work of any magnitude or complexity – in these cases bills of quantities should be used. However, for small works, drawings and a specification may well prove to be appropriate by reason of their simplicity and the direct relevance of the descriptive nature of the specification to the work to be done. This is particularly so where small-scale alterations and conversion work are being carried out, or in maintenance contracts.

Normal specification-based contracts

A specification, when used without bills of quantities, has three purposes:

(1)   It becomes the tendering document and, therefore, must be in a form which allows the estimator to price it easily. This means that

trades must be separated where possible, as this is likely to be the basis of contractors' arrangements with sub-contractors.

(2) It will, on acceptance of the tender, become a contract document and therefore it must stand up without ambiguity as part of a legal contract.

(3) It will be a management document. As well as telling the contractor exactly what work is physically to be carried out on site, it will serve the architect or quantity surveyor as a cost control document, so that valuations for interim certificates and variations can be properly ascertained.

Some measurement is still possible and desirable, but it will not have the same status as when a bill of quantities and a Standard Form of Contract with Quantities are used.

Performance specifications

Performance specifications may be used on jobs of any size and may be combined with, or form part of bills of quantities, the extent of the work being indicated by the drawings.

A performance specification specifies the performance required for the particular section or element of the work to which it relates and contractors are permitted to build to their own design and in materials of their choice, provided only that the defined performance is achieved.

If a performance specification is used for competitive tendering it is likely that the various contractors will come up with different proposals, and these have to be evaluated. Great care has to be taken in drafting the requirements of the performance specification as, if they are too rigid, little is left to the initiative of the contractor. Alternatively, if they are too loose, the evaluation becomes more difficult, as the proposed design solutions will vary more widely. Economic criteria must be taken into account when formulating the specification, otherwise the cost may be too high and the specification may have to be amended by negotiation if the whole exercise is not to be abortive.

Although a performance specification does not include a full measurement of the quantities, it is important to include a degree of measurement for the purpose of cost control. Whereas independently designed work can be measured in terms of the work to be carried out, in the case of performance specifications, the measurement has to be in terms of the function of the specification. Thus, an independently designed suspended floor will be measured in terms of the concrete, reinforcement and formwork required, together with the beams holding it up, but the floor designed to a performance specification will be measured only in terms of the total floor area to be provided. No separate measurement can be made for the beams supporting it, although there will be separate quantities whenever the performance requirements change or wherever the physical requirements of the building clearly indicate a change in the construction of the floor.

It is important that cost control should be available to the client when performance specifications are used. The best way to ensure this is to separate the various sections of work into the smallest convenient functions, so that the price for each function is isolated. In this way a substantial variation to a particular section can be more easily identified later. Thus, if the electrical installation is to be tendered for on the basis of a performance specification, it is desirable to keep the wiring for the rising main and intake separate from the distribution mains, and separate again from the wiring to outlet points in various sections of the building.

There is no reason why, following tenders on a performance specification and a particular contractor having been selected, the measurement for his particular solution to the design problem should not be made and incorporated in the contract documents for the purpose of cost control. This should be stated as a condition of the tender, so that there is no objection later.

Quantities

One of the most significant factors to affect tender and contract documentation is the decision on whether or not the work should be measured in bills of quantities. If the work is not to be quantified, the written part of the contract documentation must be provided by other means. Usually the specification becomes a contract document in its own right (either as a traditional specification or a performance specification) and this document, or a further document, then has to contain the administrative details usually contained in the preliminaries section of the bills of quantities.

The decision whether or not to measure bills of quantities may be evaluated in relation to the criteria set out below.

Criteria for measuring quantities

It might be said that measurement should be accepted as the rule on account of the major advantages set out below – and only where it can be shown positively that there is a case for not having quantities, should they be dispensed with.

The advantages of measuring bills of quantities are as follows:

- Contractors' estimating risks are considerably lower.
- The competition is much fairer, as all contractors are tendering on the same basis.
- Contractors' overheads in tendering are reduced and therefore the whole building industry works more economically.
- The measurement of quantities allows the total price to be analysed in great detail, thus providing cost feedback on the job, which in turn can be used statistically in cost planning other work.
- Bills of quantities provide the best means of controlling the cost of variations on the contract.
- Although neither the Standard Form of Contract nor the Standard Method of Measurement dictate that quantities will be used for management purposes, they do in fact provide documents which are a great help on the site, particularly for ordering and the management of sub-contract work. It is likely that new methods of measurement, coupled with computer technolgy, will lead to their more extensive use for management purposes in the future.
- The process of measuring quantities before tender is a useful test as to whether what has been drawn and specified can in fact be built. Having someone from a different profession examine the drawings and analyse construction in detail is undoubtedly helpful in identifying problems not apparent at first sight.

There are, however, also some disadvantages:

- The process of measurement is time consuming and the start on site may be delayed. This delay is to some extent offset by the saving in the tender period, because the tenderers do not have to prepare their own quantities, however abbreviated they might be. Overall, therefore, it has to be accepted that the measurement of quantities extends the procurement period – it is a matter of judgement as to whether this delay is outweighed by the other advantages.
- For the work to be measured, the design must be substantially complete – this is a requirement of the Standard Method of Measurement. The necessary timescale must therefore be taken into account – again a possible source of delay.
- Measurement of the work implies that the design is complete. There are dangers if measurement is in fact partially speculative, because the design is incomplete, but sufficient to allow a tender to be obtained and a contract to be signed before the detailed design work has been properly completed. All too often, in both private and public

sectors, the pressure to sign a contract by a certain date militates against proper completion of the detailed design work. The use of speculative measurement as a means of overcoming this difficulty may place both parties at risk and result in delay, extra cost and even dispute.

- Where the contractor has a contribution to make to the design, measurement at too early a stage may prove abortive, because some measurement may have to be amended, as the design is completed or amended, e.g. as in a two-stage tendering situation.
- Quantity surveyors' fees add to the overall cost of the project – this allegation is usually refuted on the basis that quantified tenders are more keenly priced and that post contract cost control is much more effective. It is therefore claimed that the measurement of quantities actually reduces overall project costs, even when taking fees into account.
- For small or very simple jobs, measuring quantities can be tedious and expensive in relation to the size of the job and the benefits obtained.

On balance, the advantages of the full measurement of quantities are generally overwhelming. It is likely, therefore, that this method will continue to be used in the majority of cases, because of the formidable advantages it offers, particularly in terms of financial control.

Where circumstances make the measurement of quantities inappropriate, alternative arrangements have to be made for cost control. At the very least, a detailed analysis should be made of the tender price, either with the original submission, or before work starts on site – this can be used for interim payments and for the valuation of variations. The production of this analysis should be made a condition of the tender. Alternatively, measurement can take place after tendering, in agreement with the contractor. Again this should be made a condition of the tender, so as to avoid dispute later.

This emphasis on the measurement of quantities is not to deny the importance of the specification. In terms of the quality of the work to be built, the specification is paramount. However, in terms, of the proper financial control of a contract, measurement of what is to be built is advantageous.

Bills of quantities

Pre-Contract Practice and *Contract Administration* deal with the normal situation with regard to bills of quantities. As far as tendering procedures and contractual arrangements are concerned, the principles are the same whatever layout is used for the bills. Trade layout is the easiest on which to estimate, while elemental bills are more difficult for estimating, but provide a ready basis for cost analysis. Operational bills, because they are production orientated, should allow more accurate estimating and are the most useful for management purposes during the contract. Annotated bills may be based on any of these three formats – the basic information being augmented by the addition of location and specification notes, which can be useful for site management purposes, as well as at estimating stage.

Approximate quantities

While accurate quantities are to be preferred, it may not always be possible to provide them, as noted above. In these circumstances, consideration can be given to using approximate quantities. Although not providing all the advantages described above, approximate quantities can provide a worthwhile half-way house. Circumstances in which it can be shown that approximate quantities are advantageous include the following:

- Where speed is of paramount importance and the general design is established, it may be necessary to select a contractor before production drawings can be completed. In such cases, sufficient is known of the design for approximate quantities to be produced on which a tender and a sound contract can be based.
- For groundworks, the geological conditions may not be known in sufficient detail for accurate quantities to be measured until the work is carried out. The presence or extent of soft spots or water may not be known until encountered once the work has commenced. So, although the construction may be known, the actual extent of the work required, including the general depth of excavation, may not be able to be defined at the pre-tender measurement stage and will have to be based on a broad assessment, as approximate quantities.
- In accurate bills of quantities, there will be what are described as 'approximate quantities' which will have to be remeasured when the work is actually done. Thus accurate measurement of the foundations may be possible except for the underpinning of a particular wall, which has to be an approximate quantity until that wall has been opened up. Another example, is the extra for excavation in rock, where the foundations can be measured accurately but the amount of rock to be excavated will not be known until the work is carried out.

Where bills of approximate quantities are used for tendering, the procedures will generally be exactly as with accurate bills; but the Standard Form of Contract for use With Approximate Quantities, prepared by the Joint Contracts Tribunal, should be used. The differences are, however, minor and the chief problem arises in the post contract administration stage, particularly with regard to cost control. There are three main choices available at this stage:

(1) The work can be remeasured at the end of the contract and this then becomes the basis for the final account. In this case, cost control and interim valuations become more difficult and a good deal of interim approximate measurement is necessary if these functions are to be properly performed.

(2) The work can be measured as soon as the production drawings are available, and substitution bills prepared in place of the approximate bills for the administration of the contract. This is the best method for control purposes, but undoubtedly means more work as there are then likely to be further variations on the production drawings which, in turn, entail a final account having to be measured from the substitution bills.

(3) Particularly in large contracts, where there is a resident quantity surveyor, the work may be measured immediately it is executed. This keeps the measurement up-to-date but it is of more limited value in forecasting future expenditure.

Bills of approximate quantities are obviously less useful than accurate bills and generally cannot be used for management purposes. In such cases there is little point in annotating them, as the work that they describe is to some extent hypothetical.

Schedules of rates

A schedule of rates is a tabulation of work descriptions, exactly the same as in a bill of quantities, each of which is priced on a single unit quantity.

In schedule of rates contracts the actual quantities are measured after the work has been done and the tender is, therefore, only a price list. It is obvious that a schedule of rates is unsuitable for a contract where the quantity of work is likely to affect the rate. It is most suitable for jobbing work, maintenance or redecoration. One application of schedules of rates under term contracts is described in Chapter 13, but there are others.

Schedule of rates tendering can be competitive, but it is difficult to make an objective evaluation of the rates submitted, without quantities. For this reason, quantities relating to a hypothetical, but typical, workload are often compiled and priced, using each tenderer's prices, to make a comparison.

Frequently, schedule of rates tendering is done on the basis of a pre-priced schedule, allowing the contractor to add or deduct a percentage with respect to each particular trade. Such a system of tendering and contract is common in some parts of Europe. To enable realistic tenders to be submitted, it is usual to provide as much information about the likely workload as possible. Such information might include:

- a description of the building or properties in respect of which orders will be placed
- a description of the range of work that will be ordered
- the minimum and maximum value of any one order to be issued
- the approximate value of all the work to be ordered.

Cost control under schedule of rates contracts, in the normal sense, is impossible, but this is not to say that the work cannot be properly budgeted by putting predicted costs on the project before it is started. Furthermore, the costs can be accurately accounted for. In order that valuations may be accurate and up to date, the work must be measured regularly.

'... a network analysis should be prepared ...'

Chapter 6
Competition and Negotiation

This chapter considers seven main aspects of competition and negotiation:

(1) competition
(2) the relationship between competition and negotiation
(3) negotiation
(4) when to negotiate
(5) the basis for negotiation
(6) how to negotiate
(7) advantages and disadvantages.

Competition

A contract procured by competition will involve contractors submitting tenders based upon documentation common to all. The documentation, as described in Chapter 5, will invite tenders involving competition on one or a number of the following:

- price
- time
- quality
- design
- management.

Tenders for price and time are finite and bear direct comparison with other submissions. The comparison or evaluation of tenders for design, quality and management is harder and introduces opinion and judgement by the employer and his advisers before acceptance.

The common understanding of competition is usually restricted to tenders where price and time form the major ingredients of the submission.

In such cases, the Code of Procedure for Single Stage Selective

Tendering (1996), advising on the number of tenderers invited for the size of contract, should be adhered to in preparing a list of tenderers. Careful consideration should be given to the choice of tenderers, both in their ability to carry out the work in the time given or quoted and their financial stability. Early notification should be given to each tenderer of the pre-contract and projected contract programmes, and dates should be set and agreed well in advance for the issue of the tender documents and receipt of tenders.

The relationship between competition and negotiation

Most contracts involve both competition and negotiation to a certain extent, and this might be represented by the formula:

$$C + N + X = 100\%$$

C   represents the value of work in the tender submitted on a competitive basis
N   represents that which is negotiated
X   which is usually very small, represents those parts of the contract which are neither on a competitive basis nor negotiable, but subject to factors beyond the control of the parties to the contract (such as statutory authority charges or costs arising out of legislation).

It is possible to have a contract which is virtually 100% competitive and it is also possible to have one which is nearly 100% negotiated. It is, however, rare to achieve percentages this high. The majority of contracts are a mixture of both and it is worth noting that competitive contracts will frequently have the major part of the work negotiated, perhaps in the sub-contracts. Similarly, there is many a negotiated contract where the majority of the work is actually on a competitive basis. The terms therefore tend to be related to the method of selection of the main contractor rather than to a definition of the proportions of work under competition or negotiation which make up that contract.

The distinction between a competitive and a negotiated contract has nothing whatsoever to do with whether the contract is a fixed price, target cost or reimbursement form of contract. A fixed price contract might well have the contractor selected singly on a subjective basis and the price subsequently negotiated. Equally, a cost reimbursement contract, with a fixed fee for management, can be awarded on a competitive basis.

Negotiation

Negotiation involves the agreement of prices and the basis of the contract between the parties, and usually falls under two headings, depending on the timing of events:

(1) Pre-contract negotiation – based on estimated costs for work to be let on a fixed price basis, or on an 'open-book' cost reimbursement basis.
(2) Post contract negotiation – which will nearly always involve a large element of known costs and, if there are no comparable contract rates, the examination of invoices, daywork charges and the like submitted by the contractor in justifying his entitlement on variations. This can also involve an 'open-book' approach.

Negotiation is commonly understood to relate to the cost of the work, but may also cover the programme, working conditions, quality and scope of the works when appropriate.

When to negotiate

As a general principle, on grounds of accountability, one should select a contractor and negotiate only where it can be demonstrated to the employer that there is an advantage in doing so. The following are situations in which the contractor might be selected by a method other than competition and in which it may be necessary to negotiate a price:

(1) Business relationship

The employer may have a business relationship with the contractor: for example, reciprocal trading, where an employer will go out of his way to employ a contractor who trades with him even if it means single tender action and negotiation. Another example would be a joint partnership in a development company or wholly owned subsidiary. Although in these cases it may be difficult to prove to the employer the advantages of negotiation, nevertheless, taking his total interest into account, it may be desirable.

(2) Contractor-financed projects:

If the employer finds it difficult or impossible to finance the project in any other way, then it may be desirable to select a contractor who will be willing to finance the project in the short term. In this situation, however, it is essential that the cost of financing the project is separated from the normal construction costs.

(3) Continuation of contract

The employer may have let an initial contract in competition and then find that another project with a similar design comes on programme, in which case it may be easy to prove that it would be more economical to let the second contract to the contractor carrying out the first one. This option is dealt with in detail in Chapter 13.

Sometimes an employer will have such confidence in a particular contractor, because of a large amount of work that he has carried out for him, that he will wish to select this particular contractor and negotiate, even though the particular project may be quite different from previous work. Such a course may be justified but should be carefully watched as there may be a temptation for the contractor to push up the price, until a completely uneconomic situation has arisen.

There can be a learning curve in this situation. Care should be taken to ensure that a correct balance is taken, making due allowances for both increases and savings in the cost of management, supervision, purchase of materials, letting of sub-contracts and all site operations.

(4) Special circumstances

In particular areas there may be only one contractor available to do the work. In such a case it may well be sensible to accept the situation rather than import an outside contractor. An example would be fire or storm damage to a building during its construction. Loss adjusters and insurers become involved and in some cases the name of the employer could change. However, the contractor on site would need to be instructed and negotiations are really the only solution.

(5) Special expertise or equipment

A certain contractor may be the only one available with either the expertise or the special equipment to carry out the particular project. Such a situation must be carefully assessed and if it is found that this really is the case, it is better to accept it and conduct a proper professional negotiation.

(6) Tendering climate

At times when the industry is grossly overstretched, it may well pay to negotiate a particular contract. The alternative might be to receive in competition a price embodying a very heavy premium as the contractor did not particularly want the job. A contractor will often be prepared to negotiate a reasonable price at such times, because he places value on the goodwill of a particular employer. Care must be exercised in the choice of contractor with whom to negotiate.

(7) Quick start

If a very rapid start is required, it may be necessary to negotiate. This situation must be looked at carefully, however, as it is possible to select a contractor in competition in a very short time indeed. Where a contractor has gone into liquidation, it may be an advantage to negotiate with another contractor. In this case all the design work will have been completed and the savings of time may justify selecting a single contractor and negotiating.

Other financial considerations may be far greater than the difference between negotiation and competition. An example would be a high consequential loss situation arising from material damage to a building from fire.

Generally

The above list has given an indication of the situations where the total contract might be negotiated with a single contractor. There are many other cases, however, where the contractor is selected in competition but where, because of his early selection for some particular reason, a considerable amount of negotiation of the contract price will occur (as in two-stage tendering). Such negotiation will be on the basis defined in the competitive bid (the first stage), rather than a negotiation from scratch.

Finally, there is some negotiation in nearly every contract. For example, even on a fixed price traditional form of contract with bills of quantities, there will be some negotiation in settling the final account. Although the Standard Form of Contract gives the quantity surveyor the absolute right and duty to determine a fair price for the variation, nevertheless, in practice, this is frequently negotiated and agreed with the contractor.

The basis for negotiation

There are three bases for negotiation:

(1) based on the competitive price for similar work under similar conditions in another or previous contract
(2) an agreed assessment of the estimated cost to which will be added a percentage for head office overheads and profit
(3) the actual cost of the work to which will be added a negotiated percentage for head office overheads and profit. This is best described as an 'open-book' approach.

The *first* method has certain advantages in respect of accountability. For example, it can be explained to shareholders, or an overseeing Ministry, that although the contract was negotiated, the negotiation was based on another contract which had been competitive. A 'nominated bill' may be identified which the contractor has used on another contract, not necessarily with the same client, and which then forms the basis of the negotiations.

There are, however, great dangers in this method. The other job may not form a suitable basis for all sorts of reasons, some of which may not become apparent until after negotiation has gone quite a long way. If such a method is adopted, it is essential that the employer's quantity surveyor knows as much as the contractor about the circumstances of the other job and this, by its very nature, is difficult. For example, last-minute changes in the original competitive tender may have seriously upset some of the individual prices, although such distortions may not have upset the original contract, particularly if there were no variations on those items. However, on the new contract, with different quantities, an unfair total price may result.

Where possible, the evidence of actual costs from the final account of the previous contract should be examined and brought into negotiations. Particular attention should be given to obtaining up-to-date quotations for specialist items and nominated work. Also, where the contractor intends to sub-let trades, then where time allows, competitive tenders should be obtained from three or four firms. There is a time limit to the use of historical cost or price information. It is important to review all aspects of the pricing in an endeavour to obtain a fair result for both the contractor and the employer.

The *second* method, that of assessing the estimated cost of the job, is logical because this is the normal method of tendering on a competitive basis. In this situation, the price is built up from the suppliers' and sub-contractors' quotations, together with an assessment of labour and plant requirements. Finally, after site management costs have been estimated, a percentage figure is added for head office overheads and profit. Both the client's and the contractor's quantity surveyors will participate in producing the price on this basis, which may well be documented in a normal bill of quantities.

The dangers in this method are that the estimate may be conservative and not so keen as in a competitive situation. Risks which the contractor would be prepared to take in a competitive situation, to get the job, may be exaggerated or indeed removed in the negotiated situation. The best way of getting over this is to use the assessment of estimated costs as the basis, but always to check each result in a competitive light.

The *third* method, of dealing with matters on an 'open-book' basis, where all legitimate costs to the contractor are admitted, together with a negotiated percentage for head office overheads and profit, is becoming increasingly popular. However, for this method to be successful it requires the contractor to be both organised and fair in his 'open-book' submissions and for the work to be carried out in the most efficient fashion. The risk to the employer is that the contractor may fail to achieve these requirements, at which stage he is unlikely to get the best financial deal and the project may degenerate into a dispute. To function properly this method requires total fairness and trust between all parties.

How to negotiate

Whichever basis is decided upon for the negotiation, there are certain fundamental principles that should be established as a prerequisite to negotiation.

(1) Equality of negotiators

In negotiations it is both courteous and practical to ensure a reasonable equality of status between the parties and a fair balance of numbers. It may not be sensible of course to expect the managing director of a major contracting firm to attend a meeting with the principal of a small architectural practice and his quantity surveyor to discuss only a modest sized contract. In such a case the contractor will be represented by one or two seniors from the contract management and surveying departments. At no time should meetings be arranged in which the senior level on one side tries to obtain agreement from a lower level on the other. It is important that in managing negotiations, people's duties and delegated powers are understood and defined and meetings arranged that are balanced and therefore productive.

(2) Parity of information

This should be established at the outset. Much of the information will come from the contractor's sources but if he is not prepared to share this with the employer's surveyor, suspicion will immediately result. The negotiation will take much longer as the employer's surveyor will have to get his own information. One would immediately question the motives of a contractor who was not prepared freely to give the other side the information on which the estimate was being based. This applies particularly to competitive quotations for materials and plant.

The surveyor, on his side, must respect the complete confidentiality of information passed to him. There is no doubt that parity of information is the best basis on which to build up mutual trust between the two sides, and this is the most likely way to get a successful negotiation.

Where an 'open-book' basis has been agreed upon, it is imperative that the procedure for determining the 'open-book' cost is clearly set down and agreed prior to the contract being let.

(3) Structure of estimate

The basis of negotiation must be decided from the outset. If it is to be a nominated bill, that bill must be analysed and broken down to the original basis on which the estimate was made. Only thus can variations relating to the new job become apparent. If, however, the estimate is built up from an assessment of the estimated costs, then the structure of the estimate must be agreed initially. Agreement should be reached at this stage on:

- where the head office overheads and profit are to be priced
- whether the plant is to be priced in the trades or in preliminaries
- how the labour on-costs are to be established
- the relationship between working and managing foremen
- firm price element
- risk.

Contractors have many different ways of structuring their estimates and there is no need to insist on any particular method, as long as it is clear how the estimate is structured so that each element can be identified properly.

(4) Apportionment of cost:

It is often useful to establish at an early stage an indication of where the cost is likely to lie on the project. An example might be as shown in Table 2 and Figure 4.

If such a breakdown is made it will be seen, as in this particular

Table 2 Estimated cost apportionment

Element	%
Site management	7.5
Contractor's own labour	15
Plant	5
Direct materials	20
Contractor's own sub-contractors	10
Nominated sub-contractors	25
Nominated suppliers	5
Provisional sums and contingencies	5
Head office overheads and profit	7.5
Total	100

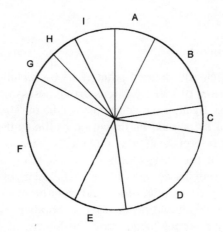

Assessed & agreed
A 7.5% site management.
B 15.0% contractor's own labour
C 5.0% plant

Competitive quotations
D 20.0% direct materials.
E 10.0% contractor's own sub-contractors.

Not negotiated at start
F 25.0% nominated sub-contractors.
G 5.0% nominated suppliers.
H 5.0% provisional sums & contingencies.

Negotiated
I 7.5% head office overheads & profit.

Figure 4. Apportionment of cost.

example, that some 35% of the contract is not to be negotiated at all at this stage. Furthermore, another 30% will be covered by competitive quotations (materials and contractor's own sub-contractors). This, therefore, leaves some 27.5% to be assessed and agreed, with 7.5% (the head office overheads and profit) to be negotiated (see Figure 4). Different jobs will, of course, have different percentages; the purpose here is only to put the negotiation into its proper perspective.

Before examining the above items individually consideration should be given to the period of the contract, on which so many of the items depend. Ideally a network analysis should be prepared and the critical path found and examined to establish the time required and where savings in time can be effective. Consideration must be given to the optimum time for construction, always bearing in mind the development as a whole. As so many of the items relating to site management and plant are directly related to time, careful consideration must be given to this aspect.

Detailed consideration should be given to all materials and sub-trades where there is evidence of a prolonged period from order to site operation. For a negotiation to be successful it should be right in price, quality and time. Delays on the contract period can result in loss and expense claims which could materially invalidate the negotiated contract sum.

(a) Site management

Having previously clearly defined what is to be included under this heading, it is a matter of making a reasonable assessment of what is going to be required. By far the biggest cost will be the site management staff and a detailed list should be drawn up of the staff required. It is desirable here that reference should be made to competitive estimates, and reasons given if the assessments of site staff on the negotiated project differ from a similar competitive one. The surveyor will also have much information from his own office and from other negotiated contracts and, whereas he should never disclose another contractor's figures, it is perfectly in order to question the figures of the particular contractor with whom he is negotiating, on the basis of his background experience.

(b) Contractor's own labour

This is the most difficult part of the assessment and the best check is to establish net prices for the work measured in position and then, by deducting the material content, to establish whether the labour is reasonable for each section of work. It must be remembered that an assessment is clearly needed for the total costs of the labour in order to establish the direct on-costs related to labour.

(c) Plant

This will either be priced in relation to the particular work it has to do (for example, excavation) or else it will be general plant such as tower cranes for which an estimate must be made of the period for which they are required. Small tools may come under this heading or may be expressed as a percentage of labour.

(d) Direct materials

These will, in most cases, be based on competitive quotations but an estimation has to be made of waste and the risk element – for example, faulty materials where the faults are disputed.

(e) Contractor's own sub-contractors

Usually these will be on the basis of competitive quotations but it may be necessary to take in subsidiary attendance in order to establish the complete cost here.

One problem that may arise concerns the amount of the sub-contractor's quotations, of which the lowest may still be considered too high. In a competitive contract the contractor will shop around after he has been awarded the contract, to get in lower quotations. In such a case it may be desirable to accept the lowest available quotation for the purpose of the tender, but to agree to substitute a lower quotation if one is later deemed advantageous so that the employer gets the benefit of the shopping around. If time allows, however, it is always better to get competitive quotations before finalising the contract.

(f) and (g) Nominated sub-contractors and nominated suppliers

Only attendance will have to be estimated here. It is assumed that the profit on them will be included in the overall profit for the job.

(h) Provisional sums and contingencies

These are part of the client's budget put into the estimate and therefore there is nothing to negotiate.

(i) Head office overheads and profit

These should be shown separately – the head office overheads, which ought to be a known factor with any contractor and therefore easily substantiated, and profit, which must provide for:

- the return on capital employed
- the risk involved in carrying out work at the estimated cost
- the market factor which could be a plus or minus quantity.

Profit is normally expressed in terms of a percentage on turnover and is therefore usually a low figure. Possibly as fair a way as any of dealing with this is to take the overall percentage profit on turnover that the contractor has achieved in his previous trading year, on the grounds that it should not be more than this when the contract is negotiated. The corresponding argument from the contractor's side would be that it should not be less because it is negotiated.

Advantages and disadvantages

Competition

The advantages of competition are:

- accountability
- a true reflection of current market prices
- quotations which can embrace comparison of contract period, design and management.

The disadvantages of competition are:

- a waste of resources in preparing several unsuccessful tenders
- restriction on full communication between design team and tenderers on details and site operations
- restriction on direct comparison of pricing with cost plan items during the tender period, resulting in necessary revisions of design at a later stage.
- a competitive tender in a depressed market may lead to the contractor receiving an unrealistically low or negative return. This could lead to the contractor attempting to cut corners on quality or getting into financial difficulties.

Negotiation

The advantages of negotiation are:

- greater speed in appointing a contractor
- full communication by all parties on design, site operations and pricing of specialist items

- progressive agreement of elemental sections of the work which might enable an earlier start on site
- creation of a team spirit between professionals and contractor and elimination of many claim items during the contract period and in the final account.

The disadvantages of negotiation are:

- a possible increase over competitive prices
- loss of time if the negotiations prove unsuccessful
- lack of accountability.

... considering the many combinations ...

Chapter 7
Fixed Price and Cost Reimbursement

Fixed price and cost reimbursement are terms used for convenience to define the two main methods of making payment. However, their use is misleading because although they indicate the only two methods available for paying for work done, very rarely is any contract discharged entirely by one or other method. Usually a combination of both methods is employed and the contract is often named according to that which predominates.

The principles, however, remain valid. Fixed price items may be defined as items paid for on the basis of a predetermined estimate of the cost of the work, including an allowance for the risk involved and the market situation in relation to the contractor's workload. The estimated price is paid by the client, irrespective of the cost incurred by the builder.

Cost reimbursement items may be defined as items paid for on the basis of the actual cost of the work.

The JCT Standard Form of Prime Cost Contract (1992) for building work offers a detailed definition of actual cost to the builder under the heading 'Definition of Prime Cost' in the second schedule.

Fixed price

It will be evident, from the definition of a fixed price item, that fixed price can apply to a unit rate, a section of work or equally to a complete contract. Similarly, it must be appreciated that a contract may consist of a multiplicity of unit rates, a series of elemental or trade sections or a single lump sum. Understanding this should dispose of the common misconception that a fixed price contract is a lump sum contract.

This, of course, is not so and indeed a fixed price contract need not have a finite sum attached to it at the beginning of the contract. A schedule of fixed rates (without quantities) is a fixed price contract, because the basis of payment has been predetermined – the price being fixed, only the quantity of work is unknown and this is ascertained by measurement as the work is done. This is still true even if the rate varies with the quantity done or used.

Cost reimbursement

In this system, payment is not based on a predetermined contractual estimate of cost. The contractor is paid whatever the work actually costs him within the limits of the contractual arrangement, which will lay down strict rules or formulae for the ascertainment or calculation of that cost.

The essential difference between fixed price and cost reimbursement is that, in the case of the fixed price contract, the contractor undertakes to do the work at a price he has estimated in advance. If he is incorrect in his estimation, then he takes the risk for being wrong. In a cost reimbursement contract, however, the employer pays the actual cost to the contractor, often described as an 'open-book' approach. If that cost is higher or lower than any estimates which may have been given for the project before it was started, then the employer automatically pays for the extra or gains from the savings.

Application to contract elements

Having established the principles, it is now possible to consider, in practical terms, the differing elements that make up a contract and how the fixed price or cost reimbursement principles apply in each case.

Unless the contract is for a single lump sum figure to supply a specified building, a 'fixed price contract' is usually made up of several elements, as follows:

- preliminaries – often a series of individual sums to represent parts of the project, such as site management costs, plant, etc.
- unit rates – work fixed in place
- PC sums – nominated sub-contractors and nominated suppliers
- provisional sums – work anticipated to be required but not yet designed, including contingencies
- sums – for work executed by a local authority or by a statutory undertaker executing work solely in pursuance of its statutory obligations
- profit – sometimes included in preliminaries and unit rates but now frequently separated as a sum or percentage in the summary.

At once it will be seen that the first two items are of a fixed price nature while the third embodies the cost reimbursement principle. The nominated sub-contracts may in themselves be on a fixed price basis as between main contractor and sub-contractor but represent a reimbursement item as between the client and main contractor. Nominated suppliers are in the same category. Provisional sums are a way of expressing that part of the employer's budget, which is not fully defined, in a contract and can eventually be resolved either way, and sums for work by local authorities and statutory undertakers are frequently on a cost reinbursement basis. Profit, however included, is always on a fixed price basis.

Even in a cost reimbursement or 'cost plus' contract the allowance included for overheads and profit is committed in advance as a lump sum, or percentage of the cost spent on the labour and materials, regardless of whether the actual cost of the overheads and management is higher or lower than this. There are therefore cost reimbursement items for labour, materials and plant, but a fixed price item for management. Where a fixed lump sum fee is given for management, an estimate of cost will be made in advance, perhaps as a percentage, as a basis for calculating this fee.

In practice, as explained above, so-called reimbursement contracts frequently contain elements of fixed price and many fixed price contracts contain elements of cost reimbursement.

Fluctuations

An element of cost reimbursement that may come into otherwise fixed price contracts, concerns the payment for fluctuations in wages and materials. A distinction is frequently made here by calling a contract where fluctuations are not paid a firm price contract. When fluctuation clauses are included in the contract and an adjustment is made for the increase or decrease in the basic cost of labour and materials, that part of the contract is dealt with on a cost reimbursement basis where the payment is based on the market price fluctuation. However, when the payment for increases or decreases in labour and materials is based on a formula related to a national index, the basis is then fixed price and not cost reimbursement. The contractor in this case is paid on a basis fixed in advance where he takes the risk and gains if he buys in a cheaper market but may lose if he cannot.

Target cost contracts

Having defined fixed price and cost reimbursement and indicated that they are the two basic methods of payment, that is not the end of the matter. It is possible to combine them, not only by having part of the work paid for on one basis and part on the other, but by applying both principles to the same work. In a firm price contract, the contractor takes the risk of his estimate being wrong, while the reverse is the case with the cost reimbursement contract where the employer is at risk in relation to the final cost. If, however, it is desired that the risk should be split between the two parties, then payment can be made in such a way as to allow this. A *target cost contract* is such a method and the system is explained in detail in Chapter 10.

Use

Just as, in discussing competitive and negotiated contracts, it was suggested that negotiations should be used only where it can be shown to be more advantageous to the employer than competition, so a similar principle applies here: fixed price will, in the majority of circumstances, benefit the employer.

The appropriate allocation of risk between the employer and the contractor will determine the apportionment of fixed price and cost reimbursement in a contract. In assessing the situation there are two main considerations, namely the circumstances of the employer and of the contractor.

The employer's position

The following criteria should be considered:

- the employer's financial position
- the employer's time requirement for building work
- the employer's corporate restrictions
- the building market.

If the employer has a defined budget and adequate time for his professional advisers to obtain fixed price quotations for the work, he should pursue that route in order to minimise his risk. If the employer has only £100 000 to spend, his advisers should limit his risk, so that this figure is not exceeded. However, if time is more important than money, the employer's position may be best protected by taking more financial risk and moving towards a cost reimbursement type of contract. It is important that the employer is advised on the correct balance between the two methods of payment. Any attempt to reduce the financial risk, without providing adequate information to the contractor, will undoubtedly backfire, with the employer having to pay more as the contractor has priced for all eventualities, covering potential risk, which may not eventually arise.

The employer may have a predetermined method of procurement for building work which will override his best interests. For example a local authority may require fixed price tenders to be obtained from an accountability point of view, yet this method may preclude the authority from having the work carried out during a particular financial year and so may lose the money allocated for the scheme.

When work in the building industry is scarce contractors will inevitably be prepared to accept greater financial risk in order to obtain work, and the employer's professional consultants should advise on that basis.

The contractor's position

The following criteria should be considered:

- the contractor's financial position
- the contractor's financially acceptable risk level
- the extent to which the work is defined
- the building market.

As with the employer, who has limited financial resources, a contractor in a similar situation cannot afford to take financial risks. Therefore, where he is asked to take such a risk, he will add a sufficient amount to his fixed price to reduce or eliminate his risk. In certain situations he may not be prepared to take the risk at all and will refuse to tender for the work. On the other hand, a contractor with substantial financial resources is in a position to take on more financial risks and submit fixed price tenders without prejudicing his immediate financial stability. Obviously, he must be satisfied that the risks he is prepared to take in pricing will even themselves out over the whole range of work that he undertakes.

The financial risk level that the contractor is prepared to take is a commercial decision. Two contractors with similar financial resources may have a different tolerance level for risk. For example, one may be prepared to estimate for material and labour cost increases, whereas the other will require to be reimbursed for any such increases. This action may preclude the latter from being invited to tender.

Where it is difficult, or impossible, to define the work clearly, as in some alteration and refurbishment schemes, it will normally be beneficial for at least some of the work to be let on a cost reimbursement basis. Attempting to estimate a fixed price for work that is not properly defined will result in the items being heavily priced to cover the risk. In addition, an undefined global item will be contractually unenforceable, if it proves to be priced inadequately, and conversely the employer will find it difficult to reclaim any money if the work involved is less than the undefined item allows for.

Where new building techniques are being used, it may be in both parties' interest to have the work carried out on a cost reimbursement basis, as the true risk may not be definable. Attempting to allocate a fixed price may, in these circumstances, result in too low or too high a price, with disastrous effects all round.

Market conditions can affect a contractor's attitude to tendering. When the building market is depressed, the contractor may take the view that he is prepared to take greater risks and tender on fixed prices. In a more buoyant market, he would be more likely to look for a cost reimbursement basis.

Programme

It will be appreciated, from what has already been stated, that cost reimbursement contracts are most commonly used where time is of the essence, or where it is impractical to define the work fully before it is executed.

Figure 5 shows a diagrammatic representation of the different sequence of events between fixed price and cost reimbursement contracts. The vertical divisions of the matrix are not intended to indicate any particular time interval – they are provided purely to assist visual comparison of the alternatives. Neither is the length of the activity bars necessarily significant, except in allowing a comparison to be drawn between the two systems. In practice, the duration of any activity must be agreed between the participants, so as to achieve the employer's expectations.

However, Figure 5 indicates that a considerable time saving can be achieved using a cost reimbursement contract – not only in achieving completion of the work, but also in completing the final account. Since the cost verification process has to be commenced from the very beginning of the contract, if progress payments and cost reporting are going to be realistic, the final account should almost be complete when payment is calculated for the last period of work.

Conclusion

When there is sufficient time, it is beneficial to carry out investigative work and produce as much detailed design and specification work as

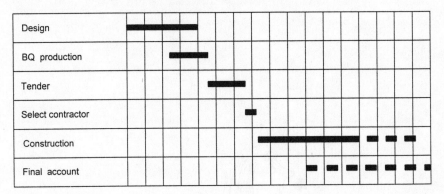

Figure 5. Sequence of events – fixed price and cost reimbursement.

possible. This work will reduce the risk element in the contract and influence the type of contract eventually used.

With the increase in use of cost reimbursement and management type contracts, often let on an 'open-book' basis, it is clear that employers are prepared to take more financial risk, in pursuit of shorter procurement times and lower prices. Partnering arrangements are often set up on an 'open-book' basis – with a view to all parties achieving best value and sharing the financial risks and benefits more equally than is the case with traditional fixed price or cost reimbursement contracts.

On all contracts, payment is made either by fixed price or cost reimbursement: FP + CR = 100%. Payment can be made in no way other than under these two principles. Almost all building contracts are a combination of both.

'... an imaginative quantity surveyor ...'

Chapter 8
Fixed Price Contracts

Having established general principles in earlier chapters, it is now appropriate to look at particular types of tendering procedure and contractual arrangement. In this chapter, fixed price contracts are examined.

In Chapter 7, fixed price was defined in terms of payment on the basis of a predetermined estimate, irrespective of the actual cost incurred by the builder. It was also established that a fixed price contract might consist of:

- a single lump sum
- a series of elemental or trade totals
- a multiplicity of unit rates.

All of these may be with or without quantities. Fixed price contracts, therefore, appear in a number of different formats, as explained below.

Historically, fixed price contracts have accounted for the majority of building contracts and they should probably still be considered as the norm, by number of contracts, if not by value. Any departure from this norm, in favour of cost reimbursement or target cost contracts, is only recommended when the circumstances specifically require a different approach in order that the best value for money can be obtained from the employer's point of view. This principle was discussed in Chapter 1, in the context of the *eternal triangle* (see Figure 1).

Most of the JCT forms of contract are essentially fixed price contracts – albeit that they usually contain optional clauses to allow an element of reimbursement of fluctuations in the prices of labour and materials. Allowing reimbursement of fluctuations passes the risk of inflation from the contractor to the employer, but other major areas of risk, such as management, supervision, resource availability and productivity, still remain the contractor's responsibility.

JCT fixed price contracts

Set out below are the JCT fixed price contracts together with the other items necessary to complete the contract documentation.

There is one other form which should be mentioned, the JCT Standard Form With Contractor's Design. This form is different from all the previous forms since its concept is based on a different premise, namely design provided by the contractor rather than by the employer's consultants. It is the only complete form produced by the JCT which recognises design by the contractor. This different concept demands different documentation. The other contract documents in this case are

Table 3 JCT Forms of Contract.

Form of contract			Other documentation
(1)	**Standard Form of Contract**		
	(a) with quantities		drawings
			bills of quantities
			(incorporating the spec)
	(b) with approximate quantities		drawings
			bills of approximate quantities
	(c) without quantities		drawings
			specification (schedule of rates for principal items to be provided after acceptance)
		optional	schedule of works
(2)	**Intermediate Form**		
	(a) with quantities		drawings
			bills of quantities
	(b) without quantities	*either*	drawings
			schedules of work (priced)
		or	drawings
			specification (priced)
		or	drawings
			specification (unpriced)
			schedule of rates/contract
			sum analysis
(3)	**Agreement for Minor Building Works**		
	(a) without quantities	*either*	drawings, with or without specification, plus schedule of work
		or	spec, with or without schedule of work
		or	schedules of work
(4)	**Standard Form of Tender & Agreement for Building Works of a Jobbing Character**		
	(a) without quantities		drawings
		and/or	specification

the Employer's Requirements, the Contractor's Proposals and a Contract Sum Analysis (see Chapter 12).

Recognising this concept of contractor's design, the JCT also provides a Contractor's Designed Portion Supplement which is used to modify the Standard Form (With or Without Quantities), to incorporate the principles identified above into the Standard Form, where part of the design is produced by consultants and part by the contractor. In this situation, the basic documentation of the Standard Form With Quantities (at 1a or 1c above) is augmented by Employer's Requirements, referring to that portion of the works designed by the contractor, and the Contractor's Proposals and Contrast Sum Analysis.

Characteristics of forms

The Standard Form of Contract

The Standard Form in its various versions, and with its Private and Local Authority formats, is the JCT's most comprehensive statement on contract conditions. The forms contain detailed conditions regulating the rights and obligations of the parties, the powers and duties of the architect and the quantity surveyor, and procedures appropriate to the variety of situations to be met on most projects of any size and complexity – even to the extent of covering the outbreak of hostilities and war damage, both of which make interesting reading.

The With Quantities, Approximate Quantities and Without Quantities versions all contain similar details and differ only in respect of those conditions referring to the contract bills, remeasurement (in the case of approximate quantities) and schedule of rates (where quantities are not measured).

This form, derived from a long list of predecessors, now runs to over 50 pages and is a complex document in itself. However, over the years a considerable body of case law has been built up which helps clarify the manner in which to interpret the document.

Depending on the amount of information available at tender stage, the appropriate version (With Quantities, Approximate Quantities, or Without Quantities) can be used. It should be noted that, with the advent of SMM7 and Co-ordinated Project Information (CPI), the choice of form used is more directly related to the information available. Bills of quantities cannot, at least in theory, be produced without the appropriate drawings and specification information being available. Before SMM7, it used to be possible for an imaginative quantity surveyor to produce an apparently comprehensive bill without the supporting drawings and details from which the contractor could built the job, thus giving a false impression of the degree of completeness of the detailed design and construction information available.

Provision is made for nominated sub-contractors and for nominated suppliers; there is a comprehensive set of tender and contract documentation available which is fully co-ordinated with the main form of contract and which ensures that the relevant conditions are fully compatible. Although the paperwork appears complicated and many employers and consultants seek to avoid the procedure for this, and other, reasons, it only formalises the basic good practice which has always been the basis of a well run contract. The documentation should always be prepared in good time and with due care.

Provision is also made for reimbursing fluctuations in the prices of labour and materials, either by traditional means or by using the formula method when quantities are provided.

Intermediate Form

This form is produced in a single format but contains a number of alternative clauses in the recitals and the appendix, which have to be selected (deleting those not required) to produce a With Quantities or Without Quantities format and to take account of a Private or Local Authority employer's requirements. The form, which is set out in a clearer format than the Standard Form, contains conditions that are less detailed than those of the Standard Forms but more detailed than those of the Agreement for Minor Building Works, and for this reason there is a note printed on the back of the form recommending its use for the middle range of contracts. This note suggests that the form would be suitable where the proposed works are:

- of simple content involving the normally recognised basic trades and skills of the industry; and
- without any building service installations of a complex nature, or other specialist work of a similar nature; and
- adequately specified, or specified and billed, as appropriate prior to the invitation of tenders.

There is also a reference to JCT Practice Note 20, in which it is suggested that the form would be most suitable where the contract period is no more than 12 months and the value not in excess of £280 000 at 1992 prices. However, there is a further rider noting that longer and larger contracts may be satisfactory if the three criteria noted above are met.

These limiting features could be overcome by incorporating special instructions elsewhere in the contract documentation, but in this situation it would be more appropriate to use the more comprehensive

Standard Form. Amending forms of contract is never recommended where it can be avoided; apart from the sheer complexity of the exercise, case law casts doubt on the effectiveness of the result. It should also be noted that JCT forms are copyright documents, which should not be amended without consent.

The provisions for the reimbursement of fluctuations in the prices of labour and materials are similar to the Standard Form, allowing for the traditional or formula methods as preferred (the formula method only being available when quantities are provided).

In the Intermediate Form an attempt is made to overcome the complexities of the nomination of sub-contractors. There are no provisions for nominating sub-contractors or suppliers, but a different and somewhat simpler concept of naming is introduced. This system allows individual sub-contractors to be selected, but once all the appropriate information is provided and a sub-contract is entered into, the specialist effectively becomes a domestic sub-contractor and the main contractor assumes total responsibility for the work (subject to various special conditions if the specialist sub-contract has to be determined). Special tender and sub-contract forms are provided to administer this process and to ensure compatibility with the main contract. It is important that this procedure is carried out in advance of the signing of the main contract, as the sub-contract must be completed within 21 days of executing the main contract.

There is no reference in this form to 'Directions' of a clerk of works and there is no provision for excluding people from the site.

Agreement for Minor Building Works

This is a single format document and is the simplest form available from the JCT for works of any substance. The form is drafted for use with drawings or specifications or a schedule of works, but no provision is made for the use of bills of quantities.

Since minor works are not considered to extend over a long period of time, no provision is made for fluctuations in the prices of labour and materials, except for tax matters imposed by government order.

Although specialist sub-contractors can be incorporated by naming them in the tender documents or in subsequent instructions, there are no provisions in the form for any special treatment or control of such specialists, nor are there any standard forms of sub-contract available for use in that situation.

Other, perhaps less significant omissions include:

- any provisions for insuring the employer's liability for damage to property other than the works, or for all-risks cover

- any provision for the ownership of materials on site to pass to the employer on payment
- any provision for opening-up and testing.

As with the Intermediate Form these limitations could be overcome by drafting special instructions, but again there seems little point in doing so when there are more comprehensive forms available covering the omissions in this very simple form.

Standard Form of Tender & Agreement for Building Works of a Jobbing Character

This form was introduced in 1990 in order to cover works of small repair and alteration, up to £10 000 in value (1990 prices), with a duration of up to one month, and a single lump sum at completion. It can also be used for a series of small jobbing contracts, in which case the conditions (JA/C 90), but not the form of tender, are used together with the employer's works orders. This form was introduced partly in order to complete the 'family' of contracts published by the JCT, covering works from the simplest to the most complex. Note that a contract administrator is not envisaged, and that the form is intended for those who are already familiar with contract procedures.

Advantages and disadvantages of fixed price contracts

The following points may be found useful in deciding to recommend to a client the use of a fixed price or cost reimbursement contract, and which of the various JCT forms of contract to use.

Advantages

- The employer avoids the risk of variations in cost due to management deficiencies, shortage of resources, reduced productivity and, to some extent, inflation, depending on which elements of the fluctuation clauses are deleted.
- Unless approximate quantities are provided, the contract sum will define the employer's financial commitment, subject to contingency and any other provisional allowances, and, of course, to any variations in the brief.
- There is an in-built incentive for the contractor to manage the work effectively and to complete the work as quickly as possible, so as to maximise his profit.

Disadvantages

- The premium paid to the contractor for taking the risk is paid irrespective of whether or not the risk materialises.
- Depending on the status of the contractor and the nature and extent of the risks involved, the premium required by the contractor may be excessive or poor value for money from the client's point of view.
- Time and production cost savings and improved productivity deriving from the learning curve are lost to the client; only the contractor benefits. This may be particularly significant on very repetitive jobs.

The choice between the various fixed price contracts available will depend on the other documentation available and the scope and complexity of the works. The advantages and disadvantages of the Standard Form, the Intermediate Form and the Minor Works Form are basically self-evident from their general characteristics set out above.

Operating procedures

The procedures for operating fixed price contracts have been set out in some detail in *Pre-Contract Practice* and in *Contract Administration*.

'... a procedure regarding checking deliveries ...'

Chapter 9
Cost Reimbursement Contracts

In Chapter 7, cost reimbursement was defined in terms of payment on the basis of the actual cost incurred by the contractor, irrespective of any estimate which may have been calculated for budgeting purposes or submitted as part of a tender. A cost reimbursement contract is therefore one in which the contractor is reimbursed his actual cost plus a fee to cover his overheads and profit.

By their very nature cost reimbursement contracts cannot have a finite sum at contract stage or when a contract is signed. The contract sum will only be ascertained at completion when the final account is settled.

Since the contractor is reimbursed his actual cost, it will be seen that, with this type of contract, the employer carries all the risk: inflation, management efficiency, effective supervision, resource availability and productivity. The contractor does take on a contractual obligation to carry out the work as economically as possible consistent with the requirements of the architect or contract administrator, having regard to the nature of the works, the prices of materials and goods and the rates of wages current at the time that the relevant work is carried out, together with other relevant circumstances. It is therefore important that the employer and his consultants have confidence in the competence of the contractor and that some method of control of the contractor's method of operation is included in the contract documentation.

The fee

The fee paid to the contractor can be a percentage figure (applied to the prime cost) or a fixed fee (a lump sum figure). A fixed fee has the advantage of providing an incentive for the contractor to work efficiently (to maximise his return). A percentage fee lacks this incentive and is sometimes considered to be an open cheque for the contractor – the more he spends, the higher his fee – and is very much akin to the concept of dayworks. Although this may be considered an extreme view, it does serve to highlight the danger of not providing an incentive for the contractor to work efficiently. However, if the scope of the work is not sufficiently defined, a fixed fee may not be acceptable to the contractor and a percentage may have to be agreed.

A refinement of the fixed fee method is to incorporate a provision for the fee to be varied as the final estimated prime cost varies in relation to the original estimated prime cost. This helps in dealing with variations.

As a further incentive to efficiency, it is possible, with either a percentage or fixed fee basis, to incorporate a provision that the fee is only adjusted if the final estimated prime cost varies from the original estimated prime cost by more than 10%, or some other agreed percentage. This adjustment could apply both to upward and downward movement.

JCT Contract Form

There is only one JCT form of contract for cost reimbursement contracts: the Prime Cost Contract. The fact that there is only one JCT form possibly indicates that only a minority of contracts are let on this basis. It also adds weight to the view, expressed in the previous chapter, that fixed price contracts may be considered to be the norm and that cost reimbursement contracts should only be used on those contracts where specific conditions render them more suitable, e.g. emergency repairs, investigatory works and such other cases where it is not possible to identify the scope of the work required to be done.

The advent of partnering arrangements has led to an increase in the use of this form of contract. If this form of contract is properly administered by both the contractor and the architect or contract administrator, the employer may benefit from not paying for risk items that do not materialise.

Characteristics of the form

The main feature distinguishing the form from that of fixed price contracts is the arrangement for payment to the contractor. Payment is based on the prime cost of the work, as defined in the documentation, and a fixed or percentage fee.

The form contains detailed conditions regulating the rights and obligations of the parties, the powers and duties of the architect and the quantity surveyor and procedures appropriate for the administration of the work. Provision is made for nominated sub-contractors and nominated suppliers and these are administered in a similar way to those under the Standard Form of Contract.

By its very essence a cost reimbursement contract does not make any provision for reimbursing fluctuations; these are dealt with automatically, as invoices and time sheets are priced at rates current at the time the work is carried out.

Advantages and disadvantages

Since fixed price contracts are considered suitable for most circumstances, there should be cogent reasons for recommending to a client the use of a cost reimbursement contract. Consideration should be given to the advantages and disadvantages of such a contract when evaluating whether or not the use would be to a client's advantage in any particular situation.

Advantages

- Work can be commenced on site immediately provided that sufficient specification information is available.
- It is often the only way of carrying out investigative work.
- It is ideal for coping with emergencies such as making buildings safe after fire damage.
- Production cost savings and improved productivity can derive from the learning curve, resulting from the repetitive nature of jobs, to the benefit of the client.
- The client pays no premium to cover risk; he pays only the actual cost of whatever is deemed to be necessary, when those risks materialise.

Disadvantages

- The client's ultimate commitment is not known.
- There is little incentive to the contractor to employ his resources efficiently (but see section above on the fee).
- The client carries nearly all the risk on the contract.
- Checking the prime cost can be a very complicated and expensive operation.

Budget and cost control

The procedure for creating a budget and controlling the cost in the early feasibility and design stages is similar to that for a normal fixed price contract. An early budget estimate should be prepared and there is no reason why cost planning should not go ahead normally. Furthermore, cost checking during the evolution of the design can also be carried out as with other types of contract.

It is important that the document showing the estimated cost of the project should be as detailed as possible so that the work is clearly defined and the basis on which the fee is calculated is established, whether it be a fixed fee or a percentage fee. It is necessary that the estimated prime cost should be divided between the work to be carried out by the contractor's own labour, the work to be carried out by his sub-contractors and the work to be carried out by nominated sub-contractors. This is fairly obvious, as the management expenses may be higher in respect of the contractor's own labour than of that of sub-contractors.

If it is decided that the fixed fee should be settled in competition and the contractor selected in that way, the detailed estimate must be sent to the tendering contractors as one of the tendering documents.

Administering the contract

The procedures for operating fixed price contracts (the norm) are dealt with in some detail in *Pre-Contract Practice* and *Contract Administration*. Here we consider only those procedures relating to cost reinbursement contracts which differ from the norm.

Although we do not normally advocate changing clauses in standard forms of contract, the Prime Cost Contract does contain a good deal of procedural matter which it is considered perfectly reasonable to vary in order that the most efficient method of working can be established on a particular project. For example, if joinery is likely to be supplied by the contractor's own joiner's shop, it may be desirable that the calculation of prime cost for that work should be on a different basis from the prime cost of work on site. It is also permissible to consider the incidence of small tools, perhaps even consumables, and to judge whether it is likely that fewer disputes will arise if some of these items are included in the fee rather than the prime cost.

There are other matters of procedure in which the quantity surveyor will be particularly interested. For instance, for cost control purposes, he may wish to ensure that the contractor sends time sheets and invoices to him at regular intervals for checking. Often a procedure for checking deliveries and delivery notes is necessary. It is desirable to ensure that the contract covers the procedures required. These are dealt with in more detail later.

Procedure for keeping prime costs

The definition of prime cost is crucial, as generally what is not specifically included in the definition is automatically deemed to be included in the fee. With small contractors, unused to working by this method, it is wise to point this out in some detail before the contract is signed, so that there is complete understanding on both sides as to what is supposed to be included in the fee. This is the best way to avoid niggling disputes.

The quantity surveyor should agree with the contractor a working method of keeping the prime cost. Contractors are sometimes inclined to take the line that as they have been chosen to carry out work on this basis, they should have a free hand and their method of keeping the prime cost should be accepted. However, a proper system of checking the prime cost is essential and does not indicate mistrust of the contractor. It is merely a prudent method of doing business and is only right and proper on accountability grounds. Furthermore, the contractor's internal method of keeping the prime cost may not, and in fact probably will not, coincide with the definition of prime cost in the contract.

The quantity surveyor will, however, be well advised to go along as far as he can with the contractor's normal costing system as this is certainly the most likely to produce error-free results. There seems to be no reason why he should not receive a copy of the weekly wage-sheet, particularly where this is prepared for the contractor in any case. If some different method is used, this must be discussed with the contractor. The quantity surveyor should not object to using another method, provided that he gets the same information as the contractor with regard to wages paid.

Labour resources

It is perhaps worthwhile at an early stage, just after the contractor is selected, to agree who is going to be the working full-time foreman to be paid as part of the prime cost, and which people are in a supervisory capacity and therefore come under the management fee.

A further fundamental point arising with cost reimbursement contracts is that, as the client is taking the risk, he should have some control over the way the work is carried out. For this reason it is desirable that the extent of the normal overtime to be worked is agreed in advance and that the contractor should then seek approval before working any further overtime. According to the type of project, thought should be given to all such matters and a decision should be made as to what can be incorporated in the documents for tendering.

Materials

A system must be worked out for the acquisition of materials. Generally speaking, competitive quotations should be obtained but it must be recognised that there may be occasions when materials are wanted so quickly that this is not possible. In such a case the quantity surveyor has to take a reasonable view and provided the materials bought are not above the market price, there is no reason why they should not be included and paid for. Even where competitive quotations can be obtained, one should consider the quality of service likely to be given to the contractor. It may be that the lowest quotation is not the most advantageous. One would, however, expect such instances to be rare.

Some system must be worked out to ensure that materials invoiced match up with materials delivered. This can be done by means of delivery tickets, and frequently contractors work such a system on their own fixed price contracts. An overall check on any particular material can be made by an assessment from the estimated prime cost. In many cases a physical check on site can easily be carried out.

Plant

The conditions of contract should state how plant is to be charged. The two main plant hire schedules in current use are:

- The RICS Schedule of Basic Plant Charges
- The Federation of Civil Engineering Contractors Daywork Schedule.

Although both these schedules are intended for use in connection with dayworks under contracts, they can, with suitable adjustment, be used for cost reimbursement contracts. Generally, it is assumed that if the plant is wanted for a long period on site, it may be cheaper to buy it outright. It may be desirable therefore that some limit upon the period of hire of plant should be laid down in the contract. Sometimes this is expressed in terms of not paying more in hire than, say, 80% of the capital cost. At the end of the contract, purchased plant can be sold or retained by the contractor, with an appropriate credit being included in the prime cost calculation.

Plant hire will invariably bring with it the question of consumable stores and discussion should take place early on how these will be treated.

Credits

A system should be established for allowing credits for surplus materials and also credits for such things as old lead taken in from existing buildings. Similarly, materials supplied by the client must be taken into account when agreeing the management fee. The sale of purchased plant has already been referred to above.

Sub-letting

On all cost reimbursement contracts, there will be a proportion of work which is sub-let on a measured basis. This may be to a sub-contractor, nominated or otherwise, or possibly to a labour only sub-contractor. Provided the arrangement is economical and competitive, there is no reason why it should not be adopted. However, if the proportion of work being sub-let rises higher than that estimated when the fee was originally fixed, there may be a case for a variation of the fee. This will have to be established before the contract is signed and, for this reason, it is desirable that the contractor's intentions regarding sub-letting on this basis should be established before he is finally selected.

Defective work

Most prime cost contracts have a clause indicating that the contractor should make good defects at his own expense. This can raise problems. If the defects are made good after practical completion there is no difficulty in separating the costs involved, but if they are made good during the progress of the work then some specific separation of the cost must be made. A kind of negative daywork charge is needed.

Cost control

Valuations for cost reimbursement contracts will normally be done on the basis of the labour paid and the invoices submitted by the contractor each month. The retention may be a percentage of the value of the work done or a proportion, say 25% of the fixed fee. There is always a considerable interval between work being carried out and invoices being submitted, and for cost control purposes, therefore, a specific reconciliation between the actual prime cost and the estimated prime cost must be made.

To do this properly, one has to go back to the detailed estimate of prime cost used for tendering and establish what proportion of that work is done. Against this, one sets the actual prime cost plus the labour and materials used but not yet invoiced. The difference between the two will show the extent to which the project is either saving on, or exceeding, the estimated prime cost. Obviously, in taking the original estimate of prime cost, allowance has to be made for any variations. From this reconciliation, estimates of the final cost of the project can be made. Although cost control with cost reimbursement contracts is more difficult than with fixed price contracts, it must be carried out and should be reasonably satisfactory.

Final account

The final account will, of course, be the total of the actual prime cost plus the management fee. It may be that during interim valuations various items are put into a suspense account pending settlement as to whether they are a proper charge against the prime cost or not. If possible, items in the suspense account should be settled and either included or rejected as soon as possible. The procedure for the final certificate, maintenance and handing over of drawings etc. in relation to the building, is similar to that of normal fixed price contracts.

'... a more sophisticated calculation ...'

Chapter 10
Target Cost Contracts

Chapter 7 examined the relationships between fixed price and cost reimbursement contracts and established the position of the target cost contract as a half-way house between fixed price and cost reimbursement contracts. Target cost contracts are a refinement of ordinary cost reimbursement contracts and introduce an element of incentive for the contractor to operate efficiently by transferring a proportion of the risk to the contractor.

The essential feature of the target cost contract is that the difference between the actual cost and the estimated cost is split in some way between the contractor and the employer. The philosophy behind this is that if the contractor cannot complete the work for the estimated cost, it is not right for the employer to pay the whole of the actual cost, as at least some of the increase may be due to the inefficiency of the contractor rather than to inaccurate estimating. On the other hand, if the contractor completes the work at a lower cost than that estimated, it may be assumed that some of the decrease is due to his own efficient management and, therefore, some of the gain should go to him. In this way the target cost contract gives the contractor a built-in incentive to manage the work as efficiently as possible.

It also, of course, gives him an incentive to increase the estimated price as much as possible in the first place. It is therefore essential that the employer takes steps to ensure that his interests are safeguarded by employing expert advice in evaluating the estimated price. This may be negotiated with the contractor before work is started or can be established by competition as part of the original tendering process.

Target cost contracts, however, should not be entered into lightly. They are expensive to manage, involving as they do both accurate measurement and careful costing on the employer's behalf. This is no doubt one reason why target cost contracts are somewhat rare; however, it is important to be aware of the availability of the system should the need arise.

In a target cost contract, payment is made partly on an estimated fixed price basis and partly on the actual prime cost. This principle can best be understood by looking at two simple examples, one showing a saving on the original target and the other showing extra:

Example 1

	£
Target (i.e. estimated total prime cost including allowance for head office overheads and profit)	100 000
Actual prime cost plus overheads etc.	90 000
Therefore, saving on target	10 000
If saving is split on a 50:50 basis, payment to contractor becomes	95 000

Example 2

Target (as before)	100 000
Actual prime cost plus overheads etc.	110 000
Therefore, deficit between actual cost and target	10 000
If deficit is split on a 50:50 basis, payment to contractor becomes	105 000

There are various points to consider in studying these examples:

(1) The figures above have been deliberately made simple; in practice the target must be revised to account for all variations. It will, therefore, be the revised target which is compared with the actual cost to establish the saving or extra.

(2) The split can be in any proportions previously agreed. Thus if it is desired that the contractor should take more of the risk, the proportions will be agreed so that the employer pays a figure nearer to the revised target and further from the actual prime cost. In the two examples given, a 25:75 split would mean that, in Example 1, the contractor would be paid £97 500, and, in Example 2, £102 500.

(3) Various methods can be used to achieve a more sophisticated calculation by splitting the work between that of the contractor and that of his own and nominated subcontractors. In addition, his head

office overheads and profit can be covered by either a fixed fee or a percentage fee. In the case of a fixed fee, it would only alter if the revised target varied from the original.

(4) The target can be obtained in various ways. The importance of reasonable accuracy has already been stressed. The best way, undoubtedly, is by means of a full and accurate bill of quantities, properly priced out and agreed between the contractor's and employer's quantity surveyors. There may, however, be circumstances when a bill of approximate quantities will suffice. For specialist work, a method of estimating in common use by the particular trade, such as a labour and material bill, may be appropriate.

In the simple examples given, prime cost is assumed to be prime cost to the employer in accordance with the detailed terms of the contract defining prime cost. For practical purposes, however, it may well be appropriate to negotiate some of the items on a fixed price basis. There are many possible variations and Example 3 (overleaf) illustrates one possible solution where three elements of the work, involving site management (5), plant (6), and overheads (7), are on a fixed price basis, while being incorporated in an overall target cost contract. Nominated sub-contractors (3) and nominated suppliers (4) are included on a full cost reimbursement basis and, for the sake of simplicity, the profit or fee (8) is taken as fixed. This leaves the labour (1) and materials (2) elements (amounting to 45% of the total) subject to the target cost calculation.

Example 3

Columns A and B represent typical overspend and underspend situations

	Target	Actual A	B
	£	£	£
(1) Prime cost of labour charges, clearly defining rates, allowances, fares, etc.	20 000	22 000	18 000
(2) Prime cost of materials and consumable stores, making allowance for discounts	25 000	26 000	24 000
(3) Nominated sub-contract work	15 000	16 000	14 000
(4) Nominated suppliers	5 000	5 000	5 000
(5) Site management (fixed)	12 000	12 000	12 000
(6) Plant, sheds, transport, etc. (fixed)	9 000	9 000	9 000
(7) Office overheads, supervision and insurance (fixed)	8 000	8 000	8 000
(8) Profit or fee (fixed)	6 000	6 000	6 000
	100 000	104 000	96 000

Column A – reduction in profit in respect of items (1) and (2) 50% of £3000	1 500
Cost to employer	102 500

Column B – additional profit in respect of items (1) and (2) 50% of £3000	1 500
Cost to employer	97 500

Guaranteed maximum price contracts

This is a type of target cost contract where the agreed guaranteed maximum price cannot be exceeded.

Any saving on the guaranteed maximum price would be shared between the contractor and the employer in a similar fashion to Column B in Example 3.

Guaranteed maximum price contracts are very difficult to set up and administer for two fundamental reasons. First, it is difficult to agree a realistic maximum price, as a contractor will inevitably try to set the figure too high, making it meaningless, and second, all variations will require the maximum price to be reassessed.

Competition

Very often the target will be negotiated between quantity surveyors on both sides. However, it is possible for the target to be the subject of a competition between contractors. This particularly applies in civil engineering works. In the latter case the bills of quantities will form the tendering document on which the target is based. Normally the contractor submitting the lowest target is selected.

From then on, the procedure is the same as with a negotiated target. The work is measured and valued on the basis of the bills, any variations being taken into account. The contractor is paid on the basis of the actual prime cost plus the relevant fee, and the difference between that figure and the measured account is shared between the contractor and the client on the lines indicated earlier.

Contract

There is no standard form of contract available for target cost work. However, the contract presents no difficulties. The JCT Prime Cost Contract can be used for all but the payments clauses. It is essential to cover a definition of what is allowable in the prime cost and this, together with the procedure for keeping the prime cost, is dealt with in Chapter 9.

Advantages and disadvantages

The advantages and disadvantages of target cost contracts are very much in line with those of cost reimbursement contracts identified in Chapter 9. The only difference, and it is an important one, is that target cost contracts have the advantage of including an incentive encouraging the contractor to operate as efficiently as possible. There is a further advantage in the flexibility of the system. The risk can be apportioned in different degrees depending on the circumstances of each party and their ability to take that risk. It need never be a black and white solution; all shades of grey are attainable according to the merits of each case.

The incorporation of a guaranteed maximum price (GMP) in a target cost contract introduces additional factors for evaluation. The advantage to the employer of the price ceiling has to be set against the disadvantage of a higher fee – for it must be appreciated that a higher premium will be charged for the additional risk taken by the contractor.

Use

The likely uses of target cost contracts have already been reviewed in Chapter 7, when considering the general use of cost reimbursement contracts. As indicated above, target cost contracts have the additional advantage of providing an incentive for the contractor to improve efficiency.

However, target cost contracts are particularly appropriate for high risk projects such as marine work where, although the extent of the work can be accurately defined, the conditions under which it is carried out may be beyond the control of both parties (for example, the incidence of tides and weather).

management contract

Chapter 11
Management and Construction Management Contracts

Management and construction management contracting are forms of contractual arrangement whereby the contractor is paid a fee to manage the building of a project on behalf of the employer. They are therefore contracts to manage, procure and supervise, rather than contracts to build.

Under management and construction management agreements, the contractor becomes a member of the employer's team – effectively the construction consultant – and it is from this factor that one of the main advantages of the arrangements is derived – the contractor working in harmony with the remainder of the construction team, instead of adversarially, as in most traditional arrangements.

While management and construction management contracts are very similar in many respects, there is one essential distinguishing characteristic which is fundamental to the understanding of the two systems. This involves the contractual arrangements between the parties. In management contracting, the works contractors, or package contractors, are in contract with the management contractor; in construction management, they are in contract with the employer. A diagrammatic representation of this difference is shown in Figure 6, below. It will be noted that, apart from this contractual link, the lines of communication are identical.

Apart from this difference, which will be referred to again later, the two systems can be evaluated together, as set out below. For convenience, the term contractor/manager is used to denote the management contractor or construction manager.

Payment and cost control

To understand the way a management or construction management contract works, it is necessary to show how payment is made.

The building work is split into sub-contracts, generally referred to as

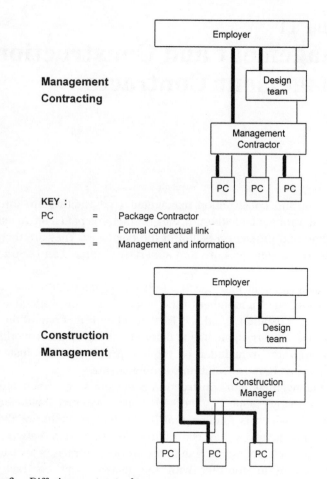

Figure 6. Differing contractual arrangements.

packages. These are normally let in competition in the usual way, although some may be negotiated or let on a cost reimbursement basis. All discounts usually revert to the client. It should be possible to arrange the various sub-contracts so that all building work is covered, but in practice a small site gang will sometimes be required to unload and help with the movement of materials on site and general cleaning after trades. Apart from this, it is unusual for the management contractor to carry out any of the building work himself, although he may have other companies within his group wishing to tender.

The contractor/manager provides the site management team and such preliminary items as offices, canteen, hoardings, etc. which are usually reimbursed at cost. On some projects, however, the cost of these forms part of his tender and is a fixed lump sum. In either case, a schedule forming part of the contract clearly defines what is required.

In addition the contractor/manager will be paid a fee. This fee is in respect of overheads (i.e. head office charges) and profit, and is usually expressed as a percentage of the final prime cost of the works. It is therefore important to distinguish the site management team from the head office staff and facilities involved. Alternatively, the fee may be based on the cost plan and not be subject to change. Thus if the cost goes up, for no good reason, the effective percentage for overheads and profit on the actual cost decreases.

Prior to the appointment of the contractor/manager, a cost plan will have been prepared. As soon as he is appointed the contractor/manager must liaise with the quantity surveyor to prepare an estimate of the prime cost showing the estimated value of all the sub-contract packages, site management and other costs. This is then agreed by the client and architect as reflecting the required level of specification and becomes the estimate of prime cost (or EPC) for the project. Valuations and certificates for payment are prepared in the usual way. Cost reporting is carried out on a package by package basis against the EPC.

Some employers require the contractor/manager to be bound by a guaranteed maximum price (or GMP), this being a sum above which any expenditure will not be reimbursed. This, however, cannot be achieved until the design and specification of the building are fairly advanced and therefore, given the nature of these contracts, cannot usually be stated in the original contract documentation.

Selection and appointment of the contractor

Selection of the contractor/manager can either be on a negotiated basis or, as is more usually the case, in competition. Negotiation is easier than for a fixed price contract, as the payment for most of the work relates to sub-contracts, which will be dealt with separately after selection of the contractor/manager. Negotiation would be on the basis of the management fee together with the contractor/manager's estimate of site management requirements.

If the contractor/manager is to be selected in competition, tender documentation should be produced inviting the submission of proposals. This documentation will advise the tenderers of everything known about the project which, given the early stage at which contractor/managers are often appointed, may not be very much but will generally include:

● general arrangement drawings
● known specification information

- the expected contract value
- details of any key dates
- details of the contract document.

The contractor/manager's submission, usually prepared in about two weeks, would normally include the following:

- the management fee
- an estimate (or tender if required) of the site management and preliminary costs
- any comments on the estimated cost of the work given the other information available
- a method statement giving outline proposals for carrying out the work, including a draft list of work packages (sub-contracts)
- a draft contract programme
- details of the proposed management team
- a proposed typical sub-contract form.

Clearly, selection will be made largely on the credibility of the contractor/manager to provide the building on time and within budget, as a difference of, say, 0.5% in the fee may not be significant in the context of some of the problems which could occur on a major development. With this in mind, the contractor/managers will be anxious to include in their submissions details of their past projects.

Some or all of the contractor/managers are generally invited to attend an interview with the employer and the design team principals, after which an appointment is made. This is usually on the basis of a Letter of Intent, which may be followed up by a Pre-Construction Agreement. The purpose of the latter is to define the responsibilities of the parties and the formula for reimbursement (usually a time charge for staff and actual expenditure) in the event that the project is aborted before commencement on site.

Contract conditions

All the management contracting and construction management firms and many clients, architects and quantity surveyors have for some years now had their own forms of contract. The JCT Form of Management Contract was issued in 1987, in an attempt to rationalise the various forms in use and establish best practice. The documentation comprises:

- the Invitation to Tender/Articles
- the Conditions

- the Employer – Works Contractor Warranty (use of which is optional)
- Phased Completion Supplements and
- a series of Practice Notes.

The use of these forms has declined with the fall in popularity of management contracting and they should now be treated with some caution.

There is no standard form of contract for construction management contracts.

Contract administration

In management contracting and construction management, it is often the case that the whole of the management team (i.e. the design team, the quantity surveyor and the contractor/manager) is based or is at least represented at senior level on site. This facilitates day-to-day communication and problem solving, consistent with the objectives of fast and flexible building. Most employers will also find it necessary to designate or appoint their own project manager, who becomes the single point of contact for the construction team.

One of the management team's first tasks is to draw up a detailed programme and procurement schedule. These show the dates by which design information is required by the quantity surveyor for the production of bills of quantities for each package. (It is usual for most packages to be sought in full competition, based on bills.) A list of possible works contractors is also compiled. These companies are then pre-qualified to ensure they are suitable, interested, financially able etc. and a number, usually four to six, are shortlisted to become the package tender list. Tenders are invited by the contractor/manager and will include full preliminaries, a works-contract programme and other details such as safety and quality control procedures. Some contractor/managers also hold mid-tender interviews to make sure the tenderers understand their part in the whole project. Any significant points arising at these meetings are, of course, circulated to all tenderers.

When the tenders are received, they are very quickly reviewed by all members of the team and one, or more, of the tenderers may be called to a post-tender interview, especially if any queries arise. A team recommendation is then made to the employer and the works contractor is appointed.

Drawings and architect's instructions are issued in the normal way. Contractor/manager's instructions are issued to the works contractors within the authority of the architect's instructions and these form the basis of the package final account. Valuations are also carried out in the normal way with works contractors' applications being checked and audited by the quantity surveyor.

Periodically, reports are made by the team to the employer, reporting on progress, on any problems being encountered or foreseen, any corrective action being taken and, of course, on the anticipated final account.

Professional advisers

Management or construction management contracts may include or exclude design work. The employer will need independent professional advice whether the design is included or not. In the case of the architect and engineers this will, of course, vary according to where the design is placed. The quantity surveyor's role is very similar to normal building contracts, but the emphasis is more on cost control and auditing of expenditure by the contractor/manager.

Advantages and disadvantages

Advantages

Among the advantages claimed for management contracting and construction management are the following:

- *Harmony, not confrontation* Traditional contracting has been described as adversarial. In management contracting and construction management, however, the sole objective of all parties is to produce the required building on time and on budget and the contractor/manager, for his part, is paid a fee for this service in a way similar to the other members of the client's team. The process should, therefore, be more harmonious.
- *Earlier start on site* Management contracting and construction management contracts allow the project to be on site earlier than by other methods. Figure 7 shows that because sub-contract tenders are invited and let whilst design is still progressing, there can be more time for each of these procedures to take place whilst allowing a completion date which is earlier than by a more traditional method.
- *Design flexibility* Management contracting and construction management contracts allow the employer and designers total

flexibility to develop the scheme, which enables each and every proposed design to be assessed in terms of the three factors affecting it, i.e. time, cost and quality. It is usual on a large project for the design team to be based on site.

- *Buildability* Early involvement of the contractor/manager means that he can advise on the suitability of proposed materials and methods with regard to time, market availability etc.
- *Early completion* Whilst the contract states a completion date, the client can, by agreement, require the works to be completed earlier, although this may attract payments for acceleration of the various work packages affected.
- *Lack of conflict* The contractor/manager has no conflict of interest. He is not prejudiced in any of the advice he gives, having nothing to gain or lose by the effects of that advice.
- *Less pricing risk* As packages are tendered closer to the time of the work being executed on site, tenderers do not have to include a premium for pricing risk.

Disadvantages

Amongst the disadvantages are the following:

- *Possible escalation of cost* Some people express concern over cost because the contractor can be appointed whilst the design is still conceptual, but in fact almost all work packages, and many of the preliminary items, should be let following competitive tenders.
- *Inflated costs* There is a tendency for contractor/managers to wish to obtain tenders only from established sub-contractors with a reliable record. This is good for ensuring satisfactory performance but if cheapness is the most important consideration to the client, regardless of standard of workmanship and finishing times, the lowest cost may not be achieved. It should also be noted that, as packages are tendered nearer to the time of the work being executed on site, the tendered cost should be more accurate and less speculative.
- *Lack of contract sum* Whilst the EPC is a carefully prepared and agreed budget which is constantly refined and in which confidence will grow as packages are let, it is not a commitment and some clients are unhappy about entering into a contract which does not have a contract sum or finite commitment.
- *Increased risk* In order to secure the contractor/manager's allegiance it is usual that he carries very little contractual risk, i.e. greater risk is carried by the client. For example, in the event of a

default by a sub-contractor, the client will usually be responsible for all the effects of this on the cost and time of the project which cannot be reclaimed from the sub-contractor.

- *Duplication* There is a possibility of duplication of some preliminary items in cases where sub-contractors are each made responsible for an item which would otherwise have been provided by the main contractor. Scaffolding is an example.

Construction Management

As noted earlier in this chapter, there is only one significant difference between management contracting and construction management – that of the parties to the package contracts being the employer and the package contractor (instead of the management contractor). This was also illustrated in Figure 6.

As already noted, none of the package contracts are entered into by the construction manager – all contracts are with the employer. This leads to some additional advantages, among which are the following:

- *Increased commitment* Whilst the construction manager employed will very likely be a company or partnership, it is the individual manager responsible who will affect the success of the project. He, not being a party to the package contracts, has virtually zero risk and therefore his sole allegiance is to the client and to the project.
- *Failure of construction manager* If, despite all the interviewing and checking, the construction manager fails to perform, a replacement can be put in place with comparative ease – certainly more easily than replacing a management contractor which is a party to numerous package contracts.

These additional advantages have led to many employers preferring construction management contracts to management contracts. There are also some additional disadvantages, among which are the following:

- *No sanctions* Since there is not usually a liquidated damages provision within the construction manager's contract of engagement, depending on the precise terms of engagement, the client may have no real sanction in the event of non-performance of the manager, short of replacing him.
- *Additional involvement* Since the employer is a party to a large number of direct contracts he is responsible, with the construction manager's assistance, for the risks associated with this and the

resolution of any disputes. For this reason this method of procurement is probably only appropriate for clients who regularly commission building work, have a measure of in-house expertise and wish to be involved in the detailed day-to-day progress of the works. However, since the employer should have a full complement of professional advisers in his team, this should not be a problem.

Use

Management or construction management contracts will very likely be appropriate in the following situations:

- where it is necessary to start work on site before the design is fully developed, usually because speed is the main priority.
- where the employer needs maximum flexibility to make changes to the building or to engage his own direct contractors during the course of construction
- where time is of the essence of the project and is of a higher priority than cost
- on large and complex projects, where the enhanced team effort pays dividends and the extra cost is comparatively insignificant in relation to the total project cost – on small projects these contracts can be very expensive.

Programme

As noted above, management and construction management contracts are often used when time is of the essence.

Figure 7 shows a diagrammatic representation of the different sequence of events between single-stage selective tendering contracts and management and construction management contracts. The vertical divisions of the matrix are not intended to indicate any particular time interval – they are provided purely to assist visual comparison of the alternatives. Neither is the length of the activity bars necessarily significant, except in allowing a comparison to be drawn between the two systems. In practice, the duration of any activity must be agreed between the participants, so as to achieve the employer's expectations.

However, Figure 7 indicates that a considerable time saving can be achieved using a management or construction management contract – although this may not be the main driving force behind the selection of this method of procurement.

Single-stage competitive tendering

Design																			
BQ production																			
Tender																			
Select contractor																			
Construction																			

Management Contracting / Construction Management

Design																			
Select manager																			
Documentation																			
Package tenders																			
Construction																			

Figure 7.   Comparative sequence of events.

Chapter 12
Design and Build Contracts

A design and build contract is a contractual arrangement in which the contractor offers to design and build a project for a sum inclusive of both design and construction costs – a single point responsibility for delivering the required project.

Design and build emerged as a method of procurement during the 1970s and 80s, offering greater confidence to clients seeking to avoid delays and exposure to costly claims and possible litigation. More recently the arrangement has been extended to include long-term maintenance and facilities management. In certain cases the contract can include the total financing of the project during its construction and fitting out. Such arrangements are often described as 'turnkey' contracts.

Contractual arrangements can vary considerably. For example, most simply, the contractor may undertake the complete design and construction, from inception to completion, using in-house designers and construction staff. Alternatively, the employer may instruct an independent design team to take the brief, prepare designs, cost the project and develop it to a point at which full planning consent can be obtained – only then entering into an agreement with the contractor who will complete the detailed design and construction. At this point, the design aspects may be taken over by the contractor's team, or the employer's team may be novated to the contractor (see 'Managing the design process' later in this chapter, which also considers the relationship of the designers to the employer and the contractor).

Design and build contracts can be let on a fixed price, or cost reimbursement, basis and can be negotiated, or subject to competitive tendering – the criteria discussed in previous chapters must be applied to the selection of the most appropriate basis. However, some element of competition is usually introduced into these contracts, and those of any substance are usually fixed price. Apart from the only JCT contract available being based on fixed price, clients usually see the establishment of an early fixed price commitment as one of the main advantages of the system.

The contract

The only complete JCT form available is the Standard Form of Building Contract With Contractor's Design 1981 Edition. However, it should be noted that the principles of this form can also be incorporated into the Standard Form of Contract, using the Contractor's Design Portion Supplement, when the contractor takes on elements of design within an otherwise consultant designed project.

This contract embodies some very different concepts to the mainstream JCT contracts for use with consultant design. The arrangements are set out in the first three recitals in the Articles of Agreement, as follows:

- *First Recital*: this identifies the works, their location and the fact that the employer has issued his requirements referred to as '**the Employer's Requirements**'.

- *Second Recital*: this confirms that the contractor has submitted proposals for carrying out the works referred to in the first recital – referred to as '**the Contractor's Proposals**' which include a statement of the sum required for carrying out the works and has also submitted an analysis of that sum – referred to as '**the Contract Sum Analysis**'.

- *Third Recital*: this records that the employer has examined the Contractor's Proposals and Contract Sum Analysis and is satisfied that they appear to meet the Employer's Requirements.

The contractor's obligation is then to design and complete the works in accordance with the details contained in the Employer's Requirements and the Contractor's Proposals. If no details are included, the contractor is free to select the materials and workmanship appropriate to the context.

It is important to note that, if there is a discrepancy between the Employer's Requirements and the Contractor's Proposals, the contract dictates that the Contractor's Proposals take precedence. The contract also covers discrepancies within the individual documents – these are usually resolved in favour of the 'innocent' party.

Under the contract, the contractor is responsible for the design and bears the same professional liability as a consultant designer. The contractor is therefore bound to exercise that reasonable care and skill expected of a competent designer. The contractor is also responsible for full compliance with statutory requirements.

The contract does not provide for the employer to appoint an architect or quantity surveyor, but instead, there is an Employer's Agent, who acts on behalf of the employer and receives or issues applications, consents, instructions, notices, requests or statements, in accordance with the conditions.

The contract includes two options for dealing with payments to the contractor and the parties should select the preferred method, and provide the appropriate information, before signing the contract. The options are:

- Stage payments (Alternative A), on application of the contractor, in accordance with the schedule of payments included in the Appendix.
- Periodic payments (Alternative B), on application by the contractor, in accordance with the period set out in the Appendix.

The stages, under Alternative A, might include an initial payment upon receipt of planning consent, if that has been the contractor's responsibility, or commencement of work on site. Under Alternative B, the first periodic payment would have to include the contractor's costs in the early stages of the process.

Where to use design and build (and when not to do so)

The manner in which design and build contracts are established can profoundly affect the quality of the final product. The recommendation to use the system must therefore be based, not only on the type of building required, but also upon the client's expectations with regards to programme, cost in construction, cost in use, level of specification and quality of design.

Design and build contracts are certainly useful for:

- Standard building types, especially for industrial or warehousing use, and particularly where early return on capital investment outweighs considerations of design excellence and capital cost.

- Buildings using proprietary systems where the manufacturer of the system might well become the main contractor; for example, repetitive housing or low cost hotels based on the assembly of factory made pods.

- Building types in which some contractors have become specialist; for example, highly serviced health care or laboratory buildings in which complexity justifies factory production.

Frequently systems of construction apply to the basic structure only, so that the choice of layout, finishes and external works can be designed to suit the client's need without financial penalty. There may well be economic advantages to the client where a contractor's proprietary system can be used without compromising the brief.

Where the Private Finance Initiative (PFI) is used as a means of procuring a building or project, bids are usually submitted by consortia, often led by the contractor, as the member of the team most likely to have the resources to fund the extensive pre-contract and tender work. Design, build, finance and operate (DBFO) schemes work in a similar way. The risk involved in speculative work on these projects is considerable, and can affect all members of the project team, although the contractor usually takes the largest share – and receives the greatest part of the reward if the bid is successful. In such circumstances the contractor will expect to enter into a design and build contract with a corporate body especially established to procure the project. The question then arises as to the nature of the relationship between the design team and the contractor, a subject considered below.

There are, however, other types of project for which design and build is a less suitable form of procurement:

- Where architectural quality is of overriding importance (the client may even wish to instigate an architectural competition).

- Where a client requires a building tailored to his special requirements.

- Where there are complex planning or environmental issues.

- Where the need for public accountability requires competitive tendering without the risk of prejudicing the quality of the works.

- Where complex refurbishment work (particularly of historic buildings) is required, as frequent or unexpected variations often arise.

- Where the brief cannot be defined, or the building function is of great complexity, so that a protracted period of research and investigation is necessary at the outset and might continue once work has commenced.

Managing the design process

The key to any design and build contract is the brief or 'client's requirements'. This may be in the form of an outline specification, a performance specification, drawings in varying degrees of detail or drawings and an outline specification in combination.

The employer often appoints a design team to assist in compiling this brief and preparing preliminary designs and costings before seeking a contractor by competition or negotiation to proceed on a design and build basis. The costing element of this exercise is important, as it is vital that the contractors are tendering on a basis that the client knows is within his budget.

It is an expensive business for design and build contractors to tender in competition, as each has to work out a detailed design to suit the brief and a price to go with it. Clearly this can be an uneconomic use of resources, which will reflect on the ultimate cost to the employer – especially if too much detail is required as part of the competition process. Before contract therefore, contractors should go no further than an outlined scheme design, together with an indicative price, based upon a specific and concise brief. Detail design, and fully detailed specification come later.

Once the contractor has been appointed and has assumed responsibility for the detailed design, the decision must be made as to the future role of the employer's consultants. When the contractor uses an in-house design team to develop the details for construction purposes, the original designers and quantity surveyor might be retained by the employer to monitor standards and supervise payments. Alternatively, the original designers might be 'novated' to the contractor, leaving the quantity surveyor to give cost advice to the client or possibly act as the employer's agent.

Novation is a legal term, used to describe a procedure whereby one contract is substituted for another – commonly, where a contract between two parties, A and B, is replaced by a contract between parties A and C. Novation should be differentiated from assignment, which is the substitution of one party for another in an existing contract. Novation is widely used on design and build contracts, when the contract between a consultant and client is replaced by a contract between the consultant and the contractor. However, as noted below, the terms and conditions of the substituted contract are, of necessity, different to some extent.

Novation will normally take place as the contract is let. The Employer's Requirements, issued as part of the tender documents, should clearly state that the consultant's appointment is to be novated, and include the precise terms and conditions of the consultant's original appointment. Since the consultant's original appointment will frequently include the provision of services which have been completed, or which are not relevant to the new appointment, the terms and conditions of the new contract are bound to be different. For example, feasibility studies and sketch proposals, forming part of the original appointment, will already have been completed and will not be required as part of the new

'Turnkey Contract'

contract, and the contractor will not necessarily require a post-contract management service. A schedule identifying the services which are to be provided under the new appointment should be drawn up and agreed by the consultant, client and contractor.

The consultant's original client will frequently be the employer under the building contract, and difficulties can arise when the client/employer fails to realise that the novated consultants no longer owe any further duty of care to the employer for that project. In practical terms the consultants are no longer employed by the employer and are no longer able to represent the employer's best interest, monitor the quality of construction or deal with payment. These duties must be left to others, possibly the employer's agent. However, depending on the extent of the design responsibility included in the contract and the contractor's own professional indemnity insurance cover, the consultants may be required to enter into collateral warranty agreements indemnifying the employer against damages arising out of design based failures.

Evaluation of submissions

When design and build tenders are returned, the evaluation of the contractor's submissions is complex, as almost certainly each contractor will have interpreted the brief in a different way, making comparison difficult. However, in most cases, the submissions can be compared on a common, or equivalent, basis, by adjusting those items which are provided in one submission but excluded from others and comparing overall value for money. In some cases, this is very simple; for example, different floor coverings can easily be costed and, if a preference is identified, adjustments can be made accordingly. However, major design differences may not be so easily evaluated or costed, except on an elemental basis.

Design qualities may be a significant factor and in the case of a design competition obviously are. In this scenario, there comes a point when the evaluation will be very subjective and cannot be expressed in money terms. Sound professional advice will then be necessary from independent architects and quantity surveyors.

After selection there will be a period during which the contractor will be refining the details of the design, to take account of the employer's final brief, preparing documentation for obtaining sub-contract tenders, and mobilising site resources. Frequently, the period of development of the design details will bring to light matters which can be incorporated in the building without penalty and to the employer's benefit. All these matters must be taken into account in finalising the formal contract documents; including the Employer's Requirements, the Contractor's

Proposals and the Contract Sum Analysis, supporting the agreed contract price.

Post-contract administration

Although the total responsibility for carrying out the design and construction work rests with the contractor, the employer is normally well advised to monitor the contractor's performance, to ensure that the specification is adhered to, as set out in the contract documents, and to check that the work is up to the agreed standard. If the original consultants' contracts have not been novated, they will be able to do this; if novation has taken place, the employer must rely on others.

Although there is no provision in the contract for this role (in terms of an architect or a clerk of works), the employer's agent has reasonable access to the works, and the employer has powers to order the opening up of covered work for testing.

Financial administration

Care must be taken to ensure that the contractor is paid the correct amount by way of interim and final payments. Again, the employer will be well advised to monitor the contractor's performance, for under this contract all the financial calculations are undertaken by the contractor. The contract does not make provision for an employer's quantity surveyor, although the employer still has the right to challenge the contractor's calculations.

The contractor is charged with making applications for payment, using either the stage payment or periodic payment system. Valuations, made by the contractor, will include the valuation of variations, which must be priced according to the valuation rules laid down in the contract. This, as with other JCT contracts, is a unilateral activity, but in this instance, of course, it is the contractor who does the calculations, and not the employer's quantity surveyor.

It is usual for the employer to appoint a quantity surveyor as his financial administrator, and preferably to retain the services of the quantity surveyor who has involved in the preparation of the brief, or employer's requirements. It is not uncommon for this appointment to coincide with that of the employer's agent.

Programme

The main driving force behind the use of design and build contracts is probably the desire to have a single point of responsibility, closely followed by the hope of a reduced risk of costs rising. However, a reduced timescale for the project is also a possibility.

Figure 8 shows a diagrammatic representation of the different sequence of events between single-stage competitive tendering and design and build contracts. The vertical devisions of the matrix are not intended to indicate any particular time interval – they are provided purely to assist visual comparison of the alternatives. Neither is the length of the activity bars necessarily significant, except in allowing a comparison to be drawn between the two systems. In practice the duration of any activity must be agreed between the participants, so as to achieve the employer's expectations.

In Figure 8, the overall project timescale is shown as the same for both systems, but as noted above, this is not intended to be significant. Depending on the size and complexity of the project, and the client's priorities, the overall timescales can be varied. However, many design and build specialists claim that programmes can be shortened, because

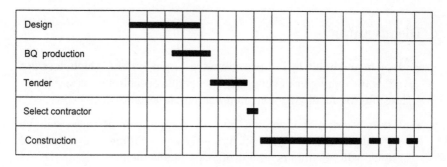

Single-stage competitive tendering

Design and Build

Figure 8. Differing processes compared.

the process is closely co-ordinated within their single responsibility and because of the shortened learning curve – the team members are used to working together.

Advantages

- Early certainty of overall contract price is obtained, as long as the JCT form of contract is used.

- Responsibility for the design, construction and performance of the building lies entirely with one party – the contractor.

- Design and build imposes a discipline, not least on the employer, to define the brief fully at an early stage. Given this, the advantages of overlapping design with construction can lead to a shorter project duration.

- Given the higher degree of co-ordination at an earlier stage, through the single point of responsibility, variations during construction tend to be fewer and the risk of post contract price escalation reduced.

Disadvantages

- Design and build, by its very nature, is a rigid system, which does not allow the employer the benefit of developing requirements and ideas, despite the facility for accommodating variations.

- Where several tenders are invited, comparison of these can be difficult, as the end product in each case is different and the final choice will be influenced by subjective judgement.

- Any variations required by the employer, after signing a contract, can prove expensive and difficult to evaluate.

- There is always a risk with regard to the quality of work. If the original brief is not precise and the specification offered by the contractor equally vague, there is a temptation for the contractor to reduce standards.

- If the design team's contract is novated to the contractor, conflict of interest can arise.

- Following novation of the design team's contract to the contractor, the employer has no direct input for checking or improving quality, unless he appoints other consultants.

- Most design and build contracts are qualified in some respects (e.g. ground conditions or the inclusion of provisional sums), which to a certain extent negates the employer's ideal of an early known financial commitment.

Postscript

In discussing the many aspects of design and build in this chapter, and especially in considering the system as an alternative to other traditional methods of procurement, the impression created, to some extent, has been one of difficulties, reduced quality and a lesser service to the client. This is partly due to the historical development of the process, but it is not necessarily the case. There are now many contractors offering an all-in service of the highest quality to clients, involving the foremost designers and producing outstanding buildings.

As always, the outcome of any procurement process depends on the quality of the input and the commitment of the parties to the process. Design and build now accounts for a sizeable proportion of the market and can be a perfectly satisfactory procurement system, as long as the parties are aware of the particular characteristics and constraints of the process.

'... small errors of documentation ...'

Chapter 13
Continuity Contracts

In Chapter 2, several aspects of the economic use of resources were considered. Among these was the subject of continuity. Continuity has two characteristics which are of interest in this context: first, economies of scale and second, the effect on the learning curve. Both characteristics have potential favourable cost implications – the first directly, in reduced material costs derived from greater buying power, and the second indirectly, through a reduction in timescales and therefore reduced labour costs.

In previous chapters, various methods of procurement and contractual arrangements have been discussed in relation to single projects or construction sites. In circumstances where continuity is available, advantage can be gained by adopting a different approach – by linking different contracts, or work packages, together.

The greatest scope for applying such concepts is obviously with clients, or agencies, that have a regular or continuing building programme. This can be for one or more new buildings or in maintenance contracts, which require the same repetitive type of work to be undertaken on a number of buildings and/or for a certain period of time.

In this chapter, the purpose and use of three types of continuity arrangements are compared and their operation is described to illustrate how benefits can be gained. The three types are:

- serial contracting
- continuation contracts
- term contracts.

Serial contracting

In many programmes of building work, such as providing several primary schools for an education authority or refurbishing a high street shopping chain for a retailer, there is an element of continuity. The object of any tendering procedure should be to ensure that building resources, including associated professional resources, are used as economically as possible. Where there is continuity these resources may be used more economically and efficiently if all the work is carried out by the same contractor, rather than by having a different one for each site or operation. It is important, therefore, to identify where the continuity lies and to evaluate the benefits that will accrue if one contractor undertakes the work instead of several.

However, it is not enough merely to establish that there is a saving in resources. As the client has provided the continuity by virtue of his programme, he should gain financially in addition to securing consistent quality and good working relationships. It will be appreciated that in any situation where one contractor is to do a series of jobs for a client, a good relationship becomes particularly important.

Serial contracting has been broadly defined as an arrangement whereby a series of contracts is let to a single contractor, but further definition is needed to distinguish it from other types of continuity arrangement. In a serial contract the approximate extent of the series is known when the offers are obtained, and the individual projects tend to be of the same order of size. The serial tender is a standing offer to carry out a series of projects on the basis of pricing information contained in competitive tendering documents. These may consist of bills of quantities, specifically designed as master bills to cover the likely items in the particular projects envisaged.

The series will usually consist of a minimum of three to a maximum of perhaps fifteen projects. The number will normally be known at the time of the tender, although further projects may be added by agreement at a later stage. Final designs will not necessarily be ready for all of them at tendering stage but the client and his architect will have a good idea of the typical requirements, and these will be reflected in the tendering documents prepared by the quantity surveyor.

Exact quantification will be achieved as the design for each project in the series is finalised. Normally a separate contract based on the original standing offer will then be negotiated and agreed for each project, based on standard items extracted from the master bills and all new or 'rogue' items will be negotiated for each contract as they occur.

Purpose and use

Serial contracts are ideally suited to a programme of work where the approximate number and size of projects to be contracted is known at the time of going to tender. For example, in refurbishing a series of high street shops for one client it will normally be the quantity of work which varies rather than the quality or specification. Similarly, although primary schools for an education authority may vary in size, in a new building programme the largest is likely to be less than three times the size of the smallest. Moreover, the sizes of most of the projects are likely to be far closer to each other than this. Therefore the same type of contractors' management is likely to be needed for them all. It is probably this latter point which should be regarded as one of the governing criteria.

Consideration should be given to the choice of this method of contracting in times of high inflation, or when there is a building boom, when it may be difficult to secure a firm price commitment over a lengthy period of time.

Operation

On serial contracts, the tendering procedure will normally be on the basis of a master bill, which will comprise most of the items envisaged in the various projects in the series. Individual projects will be controlled by having separate contracts negotiated on the basis of the master bill, after which work will proceed very much as for a normal contract.

It is desirable to take great care in the selection of the contractors for the tender list before formal tendering takes place. A mistake made when dealing with a series is likely to have more serious results than in the case of one-off contracts, so it is much more important to establish the financial and physical resources of the contractor. A bankruptcy, covering as it would many contracts rather than one, would be disastrous. It could be advantageous to make provision within the contract for the client to be able to withdraw from a contract, if performance is found to be below par. A further point to note is that small errors of documentation or detail, which may not amount to much on one contract, are clearly multiplied in a serial contract.

Comparing the contractor's capacity with the likely programme of work also requires great care. The individual contracts may be relatively small, although the total series may be large. It is possible therefore that a contractor who might not be able to cope with such a large contract on one site would be able to cope with a number of smaller projects spread over several sites, and, possibly, over a longer time. It is important, however, that his physical resources should not be stretched to the point where it is difficult for him to complete all the work. Any unprogrammed overlaps in projects could present a contractor with resourcing problems. If significant changes in the overall programme occur, it may be worth considering retendering.

It is likely that, before formal tendering, there will be a stage where interviews are held with possible tenderers, references sought, and a short-list drawn up on the basis of the information obtained.

The result of competition should be that the saving in building resources, arising from the continuity, is passed to the client. It is emphasised, however, that this is only possible where the extent of the work to be covered by the serial contract is known sufficiently for reasonable estimates of prices to be given in the original standing offer.

It should be noted that the savings derived from serial contracting are evidenced in all projects – the savings are contained in the rates in the master bill used to price all individual contracts.

Continuation contracts

A continuation contract differs from a serial contract in that it results from an *ad hoc* arrangement to take advantage of an existing situation. Thus, if a contract is already going ahead on a particular phase of a business park site, the opportunity may arise to commence a further phase with similar requirements. In this case there may have been no standing offer to do more work, the original tendering documents having been conceived only for one particular phase. However, those documents can provide a good basis for a continuation contract. Alternatively, it is possible to make provision for continuation contracts in the tendering documents for the original project, and this is some-times done. There is, however, no contractual commitment, and the possibility of a continuation contract may not arise.

In either case the contractor has to price the tender documents for the original contract on the basis of getting that particular contract only. The continuation contract, if and when it arises, is then dealt with separately – and it is only at this stage that savings arise. There is no saving on the first contract and, therefore, no benefit is derived from the potential continuity.

Purpose and use

As with serial contracts, continuation contracts are best suited to situations where the scope, style and nature of the work are very similar. The more variations there are between the projects, the more variations there will be in the contract documentation. This will result in an increased number of purely negotiated costs (rather than adjustments of competitive prices) and could also lead to organisational problems on site.

The purpose of a continuation contract should be to seek a financial advantage for the employer. Continuation contracts offer the employer the following potential benefits:

- a faster start on site, resulting from the shorter pre-contract period
- a competitive basis for pricing, resulting from the initial tender
- an experienced contractor, who has identified (and solved) the construction problems of the previous contract (this represents perhaps both a cost and a time benefit)
- re-use of the contractor's site organisation and his management team.

Whilst circumstances do arise where the use of continuation contracts might be appropriate, such as phases of a business park, the practicalities of a situation often preclude this. For example, whilst the style of the work is very similar, the projects would not necessarily be of a comparable size. Additionally uncertainties such as planning approvals, existing leases/tenancies and enabling/preparatory works often make site acquisition and site handover, ready for construction, a precarious job and all create obstacles to continuity.

Provided there is a degree of parity between the projects, however, a continuation contract may be appropriate where serial contracting is not.

Operation

In continuation contracts, the tendering documents will obviously be closely related to the documentation for the original project on which the continuation is to be based. The contract sum is negotiated on the basis of the original contract but with two substantial adjustments.

First, the contractor may require increases in the costs of labour and materials to be taken into account, to allow for the difference in time between the two tenders. The only exception to this might be where the original contract sum was on a fluctuations basis; the continuation contract could then, if necessary, be from that same base, in which case the increases would be covered entirely by the fluctuations clause. The same would apply, of course, to decreases, if any.

Second, the employer will wish the benefits of the continuity he has provided to be passed to him in the continuation contract. These will be of two kinds:

- Productivity in the construction industry generally has increased year by year and, therefore, if the continuation contract is timed, say, nine months to a year later than the original contract, there should be a saving resulting from increased productivity.

- In addition to this, however, there is a further saving arising from the economies inherent in the same contractor doing similar work for the same architect and client. This element varies from one contract to another, but may be quite substantial where similar houses or flats are being built on the second site.

The matter of measurement of productivity is difficult, and requires much research. A lot will depend on the information available to the quantity surveyor, his skill in analysing and assessing elements which affect productivity, and in negotiating with the contractor afterwards.

One of the prerequisites for the successful negotiation of a continuation contract is parity of information between client and contractor. This involves disclosing the details of how the tender for the original contract was built up. If the contractor is not prepared to do this, then it may be prudent not to proceed on this basis.

Once the continuation contract has been negotiated, a figure agreed and a contract signed, the project will proceed as any other. There is one factor, however, that should be considered, particularly where it is envisaged that several continuation contracts may arise from the original. The contractor should be given an incentive to make cost reductions. Where continuation contracts are being considered, the contractor might well be given the full benefit of any reduction in cost he makes on the first occasion, even though this involves a change in specification (but not, of course in performance), with the proviso that the full amount of the cost reduction be allowed to the client in the succeeding continuation contract. This situation arises particularly where the contractor is involved in a considerable amount of design work, and where the detailed specification is very much related to his production requirements.

Term contracts

Term contracts differ from both serial and continuation contracts in that they envisage a contractor doing certain work for a period of time or term. In this situation the contractor agrees a contract to do all work that he is asked to do within a certain framework and during a given period. Term contracts are generally set to run for a period of 12 months, but a longer period is likely to result in better tenders. Experience suggests that a term of up to three years can generally be expected to promote confidence within the period of the contract. However, the volatile nature of building costs is such that provision should be made for the employer to test the market at shorter intervals. Thus a period of two years may, on balance, be regarded as the optimum for such contracts. Provision can be made for revising unit rates regularly, relative perhaps to the annual rate of inflation.

Purpose and use

Term contracts are well suited for use with the management of large and continuing programmes of day-to-day reactive building maintenance. The management and control of such programmes can be an extremely complex task, and in view of their potential for flexibility, term contracts are perhaps best suited to the problem.

Care should be taken, both in the compilation of tendering documents, and in their evaluation. Generally speaking the tendering document will be a schedule of rates. Additionally, it can be helpful to prepare a bill of provisional quantities, based on previous years' workflow, to give an indication of the likely volume of work anticipated on specific trades over the term of the contract. As always, it is advantageous if contractors of like characteristics and performance can be selected to submit tenders, and it will usually be an advantage if multi-trade contractors can be utilised in view of the many operational, as opposed to single-trade, jobs to be found in building maintenance.

Term contracts may also be used for painting and redecoration, roadworks and other specialist trades.

JCT Contract Form

At the request of its constituent bodies, the JCT produced the Standard Form of Measured Term Contract in 1989. It is specifically intended for employers who have programmes of regular maintenance and minor works, including improvements. The JCT have also issued a Practice Note and Guide relating to the form.

This form requires the employer to list the properties in the contract area, and the type of work to be covered by the contract, as well as the term for which it is to run. The employer is required to estimate the total value of the contract and to state the maximum and minimum value of any one order. There is provision for priority coding of orders to deal with emergencies and specific programming requirements.

Payment is made on the basis of the National Schedule of Rates, or some alternative priced schedule, and the contractor quotes a single percentage adjustment to the base document, or, perhaps, different percentages for different trades. Provisions are included for fluctuations and dayworks. Measurement and valuation of individual orders can be carried out by the contract administrator or by the contractor, or can be allocated to either party according to value.

It is also possible to compile an *ad hoc* schedule, relative to any employer's particular range of work.

Operation

Term contracts are ideal for carrying out maintenance and repair work. In this type of work the individual project can be very small, say from £5, while the largest may be in the region of £30 000 or more. Tenders can be sought using a large number of methods. Two of the commonest are as follows:

(1) The employer can determine the unit or operational rate for each item and the tenderers then offer to do the work on a plus, minus or zero percentage basis to reflect their bid.
(2) Alternatively, tenders can be sought on the basis of blank schedules, which the tenderers then price.

Analysis of the tenders will usually be found to be easier if the first method is used; depending on the number of tenders received the analysis of priced schedules can be a daunting task. Experience will suggest the best method for individual circumstances. For example, a pre-priced schedule, which includes a number of blank items for multi-trade jobs, to be priced separately by the tenderer, can result in better value for money for the employer, where the price for the job is less than the sum of the component operations.

Additionally, as part of their contract, tenderers may be required to meet specified response times for various categories of work, including emergency items. The ability to meet these response times will be an important consideration in the selection of the successful contractor(s).

Contracts can be entered into with one or more contractors and the work placed on the basis of cost, committed workload and performance. Generally speaking, the more contractors available to do the work, the smoother the operation of the maintenance programme.

Orders for work are issued from time to time; the work is agreed and the contractor paid accordingly. The number of orders issued annually under a maintenance term contract can run into many thousands. To ensure the work loads can be dealt with smoothly and expeditiously, continual monitoring of the contractors' performance will be necessary and orders to the various contractors regulated accordingly.

Orders should be given a priority rating, which will require contractors to attend and carry out the work within a specified time. An incentive scheme will be found to be helpful to achieve this. Where the work is completed within a specified time, an attendance payment can be made. Alternatively, and perhaps more effectively, where the contractor fails to complete the job on time, the number of orders can be reduced until performance improves again. Additionally, a plus payment can be made for attendance to emergency call-outs. Emergencies which relate to health and safety matters can arise at any time, and will usually require quick attendance, initially to make safe only; the permanent repair being dealt with separately.

The average cost of day-to-day maintenance items will be found to be something up to £100. Assuming that the builder deals, on average, with say, 100 orders per week, his average monthly account will be in the order of £35 000 to £40 000, a not inconsiderable cashflow. It is therefore incumbent on the employer to ensure that payment of the builders' accounts is made expeditiously.

Receiving repairs requests, deciding on remedial action, determining priorities, allocating work, monitoring attendance and completion times and inspecting and certifying payment can be a complex exercise. The use of a sophisticated computer program, based on a property register, linked with a comprehensive schedule of rates, an order generator and accounts for payment will be found to be invaluable in the smooth running of a maintenance based term contract.

For an employer with a large estate and a large number of minor works of this nature to be carried out, a term contract is a method of controlling the work with a measure of accountability and a simple method of procurement for the individual project.

Partnering

Chapter 14
Partnering

In Chapter 2, partnering was identified as one of a number of ideas, or new approaches to the procurement process, in which an attempt is made to improve relationships and performance for the benefit of clients and the other members of the team – what is termed, in modern management parlance, a win–win situation. Many clients are dissatisfied with the results of individual competitively tendered projects. These contracts have often underperformed in terms of time, cost and quality and have led to adversarial attitudes from all parties.

Reference was made in the Introduction to Sir Michael Latham's Report *Constructing the Team*, Sir John Egan's report *Rethinking Construction*, the Reading Construction Forum's publication *Trusting the team; the best practice guide to partnering in construction* and the European Construction Institute's document entitled *Partnering in the public sector; a toolkit for the implementation of post award, project specific, partnering on construction projects*. These, however, are only recent manifestations of a process which has been developing for some considerable time, although the name partnering may only have been applied to it recently. It is said that the oil, chemical and power generation industries in America first used the system – while in the UK, British Airports Authority plc's framework agreements and the Bovis/Marks and Spencer set-up can be considered as forerunners to what we now recognise as partnering.

All these reports have thrown out the challenge to the industry – to improve relationships and to deliver improvements in time, quality and cost.

A definition

The terminology encountered in this field: *project partnering, strategic partnering* and *first or second generation partnering* (or even *third generation*), gives a clue to the variety of interpretations placed on the term 'partnering' – and therefore the impossibility of a finite definition.

Partnering has been described as a structured management approach which facilitates teamworking across contractual boundaries. The three fundamental characteristics, forming another version of the eternal triangle (see Figure 9), are:

- formalised mutual objectives: improved performance and reduced cost
- the active search for continuous measurable improvement
- an agreed common approach to problem resolution.

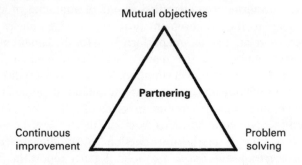

Figure 9.   The three fundamental characteristics of partnering.

Partnering should not be confused with other good project management practice, or with longstanding relationships, serial working, negotiated contracts or preferred supplier arrangements, any of which may be present, but all of which lack the structure and objective measures that support true partnering.

The essential characteristic of partnering is the commitment of all partners at all levels to make the project a success. The result is that the partnering agreement between the parties drives the relationship, rather than the contract documents. This is not to say that the contractual relationships are not important – they must be in place for the day-to-day running of the project. Standard forms of contract can be used in partnering, although it is as well to check the specific terms of any document to make sure that they do not cut across the aims of the partnering agreement.

Partnering can be applied to single jobs, often then termed *project partnering*, but the benefits of partnering tend to be cumulative, so that significantly greater benefits arise over several projects – a situation often termed *strategic partnering*. It is also considered that the benefits to the client are further enhanced if partnering is applied throughout the supply chain, thus covering the employer, the professional advisers, the contractor and sub-contractors and suppliers. If the client's workload is sufficiently large, or geographically dispersed, there is no reason why several different partnering agreements should not be set up, running concurrently, with different groups of partners.

It should be born in mind that according to EU and UK regulations for public sector procurement, partnering can only be applied to projects after the award of the contract. The European Construction Institute's document *Partnering in the public sector; a toolkit for the implementation of post award, project specific, partnering on construction projects* is particularly relevant in this context.

The key to partnering is a change in mental attitude: a new approach, based on honesty and co-operation. Without a fundamental change in attitude, incorporating these two ingredients, no partnering agreement will work, or achieve the significant improvements in performance claimed to be achievable.

The agreement

A partnering agreement may be a verbal agreement, a broad written arrangement on a single piece of paper, or a more detailed and formal document. Each situation will have different aims and requirements, which will dictate the appropriate method of setting up the partnering agreement. Care has to be taken, however, to ensure that a detailed written agreement does not lead to an adversarial attitude if things go wrong.

The intention of a partnering agreement is to establish a framework, based on mutual objectives and agreed procedures, allowing the participants to maximise specific business goals. A partnering agreement should consider the following areas:

- The formulation of a statement outlining the philosophy of the parties: that of working in good faith.
- The setting of realistic and achievable targets, with a procedure to monitor and review these as necessary.
- The agreement of a method where open-book costs can be established and savings can be shared by all participants.
- A procedural framework, including the definition of the parties' roles, responsibilities and lines of communication.
- An agreed procedure for the resolution of disputes, without damaging the original objectives of the partnering agreement.

The partnering workshop

The practical details of the partnering agreement, its procedures and objectives are often worked out in a one or two day workshop, where all the participants come together, at the earliest possible stage, to plan the project. This is the start of the teambuilding exercise and involves the actual personnel to be involved in the project, not just the senior management or marketing staff.

The workshop is often run by professional facilitators and terminates with the drawing up of the project charter, which is often signed by all participants.

The benefits

A properly set up and effective partnering agreement should seek to achieve benefits in respect of time, cost and quality. Such benefits will be derived in a number of ways, based on the following concepts:

- *Reduced learning curve*: optimum performance and productivity can be achieved much earlier in the process, due to familiarity with working practices and confidence in other participants' expertise.
- *Administrative efficiency*: continuity of personnel and familiar systems of management eliminate teething problems and silly mistakes.
- *Reduced abortive tendering costs*: overheads involved in tendering are reduced. In traditional competitive tendering, a success rate of perhaps a one in five or six is not unusual; this adds a heavy burden to overhead costs.
- *Improved communications and decision procedures*: structured procedures, developed during regular contact, promote confidence and reliability.
- *Improved quality and programming*: an overall strategy, giving a proper balance to the client's priorities, avoids the achievement of one target at the expense of others, so quality and time are not necessarily sacrificed to cost.
- *Economies of scale*: advantage can be taken of continuity and preferred supplier arrangements, so partnering exists throughout the supply chain.
- *Continuity of work and personnel*: long-term relationships can be developed, building confidence and reliability.
- *Risk*: risk identification is improved, leading to better risk allocation and management.

- *Problem solving*: problems are identified at an early stage, thus facilitating an early solution and avoiding escalation. If a formal dispute does arise, agreed procedures can be implemented quickly to achieve an early resolution.
- *Non-adversarial approach*: this avoids legal costs in most cases and minimises them in adjudication, arbitration and litigation.

The risks

Partnering arrangements are no panacea for the ills of the construction industry and the benefits do not come automatically. Partnering needs commitment at all levels in the organisations involved, if the benefits are to be fully realised. However, the parties must be vigilant, to prevent complacency setting in. Among the risks attached to partnering, like many other innovative systems, are the following:

- *Potential lack of accountability*: the absence of competition means that other mechanisms are required to demonstrate the absence of collusion; quality and audit systems and the realisation of mutual benefits should satisfy both parties.
- *Unrealistic targeting*: the search for constant improvement can lead to targets for cost, quality and time being either too low or too high; again the open-book approach to management should provide reassurance.
- *Commercial pressure*: contractors' (or their shareholders') desire to increase returns may tempt them not to declare openly the true cost savings achieved.
- *Cost attitude*: having removed, or reduced, the adversarial approach in favour of teamwork, care must be taken to avoid slackness and the taking of short cuts. As noted above, the ordinary contract rights and responsibilities must continue to be observed.
- *Change of personnel*: it is impossible to prevent some turnover of staff and there is a danger that new members of the team will not understand the concept, or not be as committed to the project as those they replace.
- *Benefits slow to materialise*: the full effect of the benefits may not appear immediately; indeed some will not materialise unless a number of projects are undertaken. Careful analysis of the potential benefits, accurate targeting and proper evaluation of the feedback are required to avoid disillusionment when results appear to be slow to materialise.
- *Programme amendments*: when the employer's requirements or

priorities change, or his programme of projects is varied, the mutual goals may no longer be achievable. It is important to analyse fully the effect of any variations in the brief on all members of the partnering arrangement.

- *Dissolution on failure*: the termination of any partnering arrangement, for whatever reason, may lead to financial problems when unravelling complex relationships.

Future of partnering

For partnering to be successful, it requires a more flexible attitude and approach than has traditionally been the case in the industry, and an acceptance that organisations and individuals are entitled to receive a reasonable return for their efforts. It also requires people to accept that the cheapest price does not necessarily provide best value for money. A solid base of traditional training and good professional and commercial practice needs to be extended within the industry to allow the more lateral approach of a partnering arrangement to benefit all concerned.

The essence of partnering is in the attitude of the participants, and in their ability to achieve the optimum overall solution, given a particular set of requirements and circumstances, while ensuring that all participants benefit.

'... incentive ...'

Appendix A
Glossary of Terms

Many terms used in tendering and contract involve words in common use in the English language which have been given a particular meaning in this context. The definitions below are deliberately kept brief, as in many cases a full discussion of the meaning is to be found in the text. It is hoped that these brief definitions will help the reader when the words are encountered before the subject has been fully discussed.

ACCOUNTABILITY
The responsibility to justify, to those on whose behalf contracts are let, the procedures and decisions for the selection and payment of contractors and for the administration of the contract.

BILLS OF APPROXIMATE QUANTITIES
As BILLS OF QUANTITIES, but the quantities are approximate – note that the descriptions of work are still accurate. Remeasurement is required to ascertain final commitment.

BILLS OF QUANTITIES
The quantified measurement and description of work in a proposed building contract – a contract document in the Standard Form. Bills of quantities usually include the specification and preliminaries (administrative and contract detail).

BUILDING RESOURCES
The physical resources needed to build, in terms of labour, materials, plant, management, consumable stores and finance.

CDM
The Construction (Design and Management) Regulations 1994, which came into effect on 31 March 1995. Regulations which place statutory duties on employers, planning supervisors, designers and contractors, to plan, co-ordinate and manage health and safety throughout all stages of a construction project.

COMMON ARRANGEMENT
A system for the co-ordination of project information based on a natural grouping of work sections within the

construction industry, to be used for the preparation of drawings, specifications and bills of quantities. It can also be used for the development of library arrangements for product information.

COMPETITION

The endeavour, simultaneously with others, to gain a contract – applicable to both main and sub-contracts and used in respect of both selection of the contractor and determination of price.

CONSIDERATION

In the law of contract, an essential in making a valid contract – some *quid pro quo* given by the promisee in return for the promise made by the promisor – it is the price for which the promise is bought – mere assent is not sufficient.

CONSTRUCTION ACT

The Housing Grants, Construction and Regeneration Act 1996, Part II (Construction Contracts) of which came into force on 1 May 1998, together with the Scheme for Construction Contracts (England and Wales) Regulations. Part II confers statutory rights on a party to a construction contract, in connection with payment, suspension of work and adjudication.

CONSTRUCTION MANAGEMENT

Contractual arrangements, similar to management contracting, except that the employer enters into contract with package contractors, rather than the contractor.

CONTINUATION CONTRACT

Contract negotiated with existing contractor to continue the involvement of the same production team as on a previous contract – with a view to obtaining cost savings from continuity. See also CONTINUITY.

CONTINUITY

The linking of otherwise separate activities or the extension of the participation of those involved in construction projects, with a view to obtaining economic advantage from the economies of scale and the reduction of the learning curve. Particularly applies to contractors who traditionally have had to tender each project separately. See also CONTINUITY CONTRACTS, SERIAL TENDERING and TERM CONTRACTS.

CONTINUITY CONTRACTS

Contracts incorporating continuity – linking otherwise separate projects for economic advantage. See CONTINUATION CONTRACTS, TERM CONTRACTS and SERIAL TENDERING.

CONTRACT DOCUMENTS	The documents signed by both parties and precisely stated in the form of contract as a 'Contract Document'. Contracts can be signed under hand or be executed as a deed.
CONTRACT PERIOD	The length of building time on site between the date for possession of the site to the date of completion stated in the contract or the certificate of practical completion.
CONTRACT SUM	The sum as defined in the contract to be paid to the contractor for work to be carried out.
CONTRACTUAL ARRANGEMENTS	All those arrangements in the contract documents showing the obligations, rights and liabilities of the parties.
COST	Strictly, the cost to the contractor as distinct from the price to the employer. However, sometimes used interchangeably with price.
COST CONTROL	In spite of the definition of cost, the term normally refers to the control of cost to the employer (i.e. price) from budget right through design and construction to final payment.
COST PLAN	A plan, normally formulated by the quantity surveyor, in which the estimated cost of a building is apportioned among the various construction elements to enable control of the cost to be exercised and to allow, within each element, comparison of various methods of construction and specification.
COST PLUS	A contractual arrangement whereby the contractor is paid the net prime cost plus a percentage or lump sum fee to cover overheads and profit. See also PRIME COST.
COST REIMBURSEMENT	Reimbursement by the employer of the net costs the contractor has paid out – one of the two fundamental methods of payment to contractors, the other being fixed price. In practice, nearly always modified to include an element of fixed price, i.e. obtaining competitive quotations for specialist sub-contract work.
DBFO	Design, Build, Finance and Operate – an extension of the PFI concept to include a defined period of operation of the completed project, during which a 'toll' or 'rent' is

charged to recoup the development cost. After the defined period of operation, the project is handed over to the public sector. See also PFI.

DEMOLITION CONTRACT

A contract for the demolition of a building, or other structure, let separately from the main building contract and normally carried out by a specialist demolition contractor.

DESIGN AND BUILD
CONTRACT

Design and build contracts are tender and contractual arrangements whereby the contractor offers to design and build a project for a sum inclusive of both the design fee and construction costs.

DETERMINATION OF PRICE

The establishment of the price to be paid for work – often bound up with selection of the contractor but not necessarily so.

ELEMENTAL BILLS

Bills of quantities presented in elemental format rather than trade by trade. The elements normally represent the function of the construction as far as practicable and will be used for cost analysis.

EXTENDED ARM
CONTRACTING

A system whereby preferred contractors (who have pre-qualified) take on the additional role of project management.

FACILITIES MANAGEMENT

A service offered to property owners, covering advice on the total management of a single building, or a complete corporate estate, including its development, fitting out, servicing and maintenance, with a view to providing best value for money in a life cycle context. See also LIFE CYCLE COSTING.

FIRM PRICE

A contractual arrangement where labour and material price fluctuations are at the entire risk or benefit of the contractor. However, changes in government taxes are usually given or paid in a firm price contract.

FIXED FEE

A fee paid generally for management, fixed in advance, where the remainder of the contract is on a cost reimbursement basis; such a contract is sometimes referred to as 'cost plus fixed fee'.

FIXED PRICE

One of the two fundamental methods of contract payment (the other being cost reimbursement), based on estimated figures established in advance of the work being carried out, which includes all the allowances required by the contractor to undertake the work including risk, factors reflecting the state of the market and the contractor's workload. May be in respect of the total work, a part, or units of work.

FLUCTUATIONS

The increases or decreases in cost of employing labour and purchasing materials – not variations which relate to changes in scope of work in a contract.

GMP

Guaranteed maximum price – an upper limit sometimes applied to target cost contracts, giving the client a guaranteed maximum commitment, while still giving the contractor an incentive to improve performance.

INCENTIVE

Motivation to achieve – used particularly in contracts in respect of the possibility of the contractor making reductions in the cost of carrying out work.

LIFE CYCLE COSTING

A technique for identifying best value for money over the whole life of a product, element or building, including its original provision and planned maintenance throughout its lifetime – its cost in use. Often leads to improved quality at inception to reduce maintenance during the subsequent lifetime, achieving best value in terms of capital and revenue cost.

LUMP SUM

A fixed price, but in a single sum for the total contract work and not intended to be adjusted by variation or remeasurement.

MANAGEMENT CONTRACTING

Management contracting is a form of contractual arrangement whereby a contractor is paid a fee to manage the building of a project on behalf of the client. It is, in essence, a contract to manage rather than a contract to build. The contractor enters into contract with package contractors in the normal way (see also CONSTRUCTION MANAGEMENT).

NEGOTIATION

The settlement of prices by agreement. Negotiated contract: a contract price based on negotiated prices instead of being awarded as a result of competition. In main contract terms, negotiation usually refers only to the main contractor's own work as sub-contract work is often awarded through competition.

NOMINATED SUB-CONTRACT
A sub-contract where the sub-contractor is nominated by the architect or engineer and not selected by the contractor.

OPERATIONAL BILLS
Bills of quantities presented in a form reflecting the likely operations on site for each section of work.

PFI
Private Finance Initiative: a system whereby the private sector (usually as a consortium) undertakes to finance the total procurement process on behalf of the public sector, payment being delayed until the project is complete and ready for occupation at handover. See also DBFO (Design Build Finance and Operate).

PARALLEL WORKING
A method of working whereby the production drawings will be produced in parallel with work on site, in accordance with an agreed programme.

PARITY OF INFORMATION
The free exchange between parties of estimating information which might be considered confidential under normal competition, but should be a condition precedent to negotiation.

PARTNERING
A special relationship developed between the two parties to a contract (but may include several partners), normally over a programme of work, rather than a single contract, whereby quality and efficiency are continuously improved and both parties derive economic advantage. Advance targets are often set as an incentive (see chapter 14 for further details).

PERFORMANCE SPECIFICATION
A specification of work related to the performance required (e.g. operation, capacity, strength, etc.) rather than in terms of the workmanship and materials to be used.

PRICE
The cost to the employer – normally in a fixed price contract will be based on the contractor's estimate of his own cost, plus what he thinks he can obtain as profit in relation to the current market situation.

PRIME COST
The net cost to the contractor of a particular item or piece of work. The initials PC used in bills of quantities and specifications have a particular contractual implication denoting work to be carried out by nominated sub-contractors or suppliers.

PRODUCTION COST SAVINGS

Savings in the contractor's costs of carrying out the work – productivity savings refer to methods of carrying out work more efficiently, particularly in respect of the use of labour and plant.

PROPRIETARY SYSTEM

A system of building designed by the contractor or sub-contractor and usually used by him alone or by others under licence.

PROVISIONAL SUM – DEFINED OR UNDEFINED WORK

A sum included in a tender or contract for work anticipated but not sufficiently designed or detailed to permit definitive description and measurement. Where the work can be adequately 'defined' as set out in SMM7, General Rule 10, the contractor must allow for the work in his programme and pricing of preliminaries. Where the work cannot be so defined, the contractor is deemed not to have allowed for it in his programme and pricing.

RISK

The hazard inherent in forecasting the cost of work to be carried out at a future date, which results in exposure to loss. Either party to a contract can agree beforehand to accept the risk of such loss for the whole or part of the work. If the contractor accepts it, he will normally add something (a premium) to allow for it. If the client accepts it, he pays the actual cost, not the estimated cost.

SMM7

Standard Method of Measurement of Building Works: Seventh Edition. A code of measurement practice agreed between the Royal Institution of Chartered Surveyors and the Building Employers Confederation, which is incorporated in all JCT 'with-quantities' contracts.

SCHEDULE OF RATES

A list of unit items of work priced at a rate per unit. The schedule thus formed is used in conjunction with the measurement of work to calculate payment.

SERIAL TENDERING

A standing offer to carry out work for more than one project in accordance with the tender submitted for the initial project, or based on hypothetical bills of quantities representing the average project of a series. A serial contract is the contract relating to such a tender.

SITE MANAGEMENT

The personnel carrying out management work full-time on site, as against head office staff, who will be involved only part-time and deal with activities for the firm as a whole.

SPECIFICATION — Unquantified description of the work required to be done. For building work under the JCT Standard Form of Contract With Quantities, the specification is incorporated in the bills of quantities and is not a separate contract document.

STANDARD FORMS OF CONTRACT — The forms of contract for building work published by the Joint Contracts Tribunal, referred to as the JCT Forms. The constituent bodies of the JCT and the types of forms available are listed in Appendix B.

SUB-CONTRACTOR (DOMESTIC) — A sub-contractor selected by the contractor, sometimes from a short list of sub-contractors listed in the tender document.

SUBSTITUTION BILLS — In contracts based on bills of approximate quantities, bills prepared as production information becomes available which are substituted for the original approximate bills.

TARGET COST CONTRACT — A contract in which the contract sum is calculated by comparing the actual prime cost with an estimate of the cost (target) made in advance, the difference between the two being shared between the parties according to the terms set out in the contract.

TENDER — An offer to carry out work at a price – a bid in accordance with the conditions set down in the tender documents.

TENDER DOCUMENTS — All documents upon which the tender is based. Quite often the tender documents are incorporated in the contract as 'contract documents'.

TENDERING PROCEDURE — The process whereby a contractor is selected to carry out work and an offer (or price) is obtained on which a contract may be let.

TERM CONTRACT — A method of carrying out work for a fixed term, on the basis of a schedule of rates – usually used for alteration and repair work by large organisations and in the public sector, where a continuous workload is anticipated.

TURNKEY CONTRACT

A contract where the contractor provides the total resources required, including the finance as well as design, construction and fitting out.

TWO-STAGE TENDERING

Commonly used to indicate situations where a competition takes place to select a contractor at an early stage on a basis which, because of the lack of information then available, precludes the determination of the contract price, this being settled later at the second stage, by negotiation.

VALUE MANAGEMENT

An extension of the concept of value engineering to cover the whole project, from the earliest feasibility stage, reviewing the brief and design solutions in order to optimise function and cost. See also VALUE ENGINEERING.

VALUE ENGINEERING

The refinement of the specification of products and processes, so as to achieve the required performance at lowest cost – by the systematic identification of the function of each product or process in its context and establishing the lowest cost while still maintaining that function. See also VALUE MANAGEMENT.

VARIATIONS

The contract term for changes in work authorised by the architect on behalf of the employer – to be distinguished from fluctuations.

Appendix B
Standard Forms of Building Contract

The most common forms in use for building work are those issued by the Joint Contracts Tribunal, whose constituent bodies, following incorporation as an independent company, are:

Association of Consulting Engineers
British Property Federation
Construction Confederation
Local Government Association
National Specialist Contractors Council
Royal Institute of British Architects
Royal Institution of Chartered Surveyors
Scottish Building Contract Committee

The Joint Contracts Tribunal have issued and regularly amend the following Standard Forms of Contract with supporting documentation and Practice Notes.

JCT Standard Form of Prime Cost Contract

JCT Standard Forms of Building Contract

Local Authorities With Quantities
Local Authorities Without Quantities
Local Authorities With Approximate Quantities
Private With Quantities
Private Without Quantities
Private With Approximate Quantities
Fluctuation Clauses (Local Authorities and Private Editions)
Formula Rules
Sectional Completion Supplement (With or Without Quantities)
Contractor's Designed Portion Supplement With Quantities
Contractor's Designed Portion Supplement Without Quantities
Nominated Sub-Contract Tender – NSC/1 or 1a
Nominated Sub-Contractor Agreement – NSC/2
Form of Nomination of Sub-Contractor – NSC/3 or 3a
Nominated Sub-Contracts – NSC/4 or 4a
Nominated Supplier Tender – TNS/1
Nominated Supplier Warranty – TNS/2

Domestic Sub-Contract Articles of Agreement – DOM/1
Domestic Sub-Contract Conditions – DOM/1

JCT Standard Form of Building Contract with Contractor's Design

Standard Form of Contract
Formula Rules
Domestic Sub-Contract Articles of Agreement – DOM/2

JCT Intermediate Form of Building Contract

Standard Form of Contract
Fluctuation Clauses
Formula Rules
Sectional Completion Supplement
Form of Tender and Agreement for Named Sub-Contractor – NAM/T
Named Sub-Contract Conditions – NAM/SC
Form of Employer/Specialist Agreement – ESA/1
Domestic Sub-Contract Articles of Agreement – IN/SC
Domestic Sub-Contract Conditions – IN/SC

JCT Standard Form of Management Contract

Standard Form of Contract
Works Contract/1 – Section 1 – Invitation to Tender
Works Contract/1 – Section 2 – Tender by the Works Contractor
Works Contract/1 – Section 3 – Articles of Agreement
Works Contract/2 – Works Contract Conditions
Works Contract/3 – Employer/Works Contractor Agreement
Phased Completion Supplements (Management Contracts and Works Contract)
Formula Rules

JCT Agreement for Minor Building Works

JCT Standard Form of Measured Term Contract

JCT Agreement for Renovation Grant Works With Architect – RG/A

JCT Agreement for Renovation Grant Works Without Architect – RG/C

JCT Jobbing Agreement

JCT Arbitration Rules (CIMAR)

Collateral Warranty Form of Agreement MCWa/F (Funder)

Collateral Warranty Form of Agreement MCWa/P & T (Purchaser or Tenant)

The following forms have been published by the Association of Consulting Architects and the British Property Federation:

ACA Form of Building Agreement, Third Edition
ACA Form of Sub-Contract, Third Edition
ACA Form of Building Agreement BPF Edition

The following forms are used for government contracts:

GC/Works/1 With Quantities (1998)
   For use on major contracts, having a contract value greater than £200 000

GC/Works/1 Without Quantities (1998)
   For use on major contracts, having a recommended contract value up to £500 000, (with a maximum value of £2 000 000)

GC/Works/1 Single Stage Design & Build (1998)
   For use with a single-stage tendering procedure, without a separate design stage

GC/Works/2 (1998)
   For use on minor building and civil engineering work, having a contract value of approximately £25 000 to £200 000), and demolitions of any size. It is for use with specification, drawings and a schedule of rates

GC/Works/3 (1998)
   For use with mechanical and electrical works of any value. It is for use with specification and drawings and the optional use of bills of quantities or a schedule of rates

GC/Works/4 (1998)
   For use on small building, civil engineering and mechanical and electrical engineering works, having a contract value up to £75 000. It is for use with specification and drawings only. There is no provision for a schedule of rates and payment is arranged by thirds of the work by value

There are also Standard Forms of Contract which are published specifically for civil engineering works. Conditions of Contract (6th Edition) and Forms of Tender, Agreement and Bond for Works of Civil Engineering Construction (commonly known as the ICE Conditions of Contract) are issued by:

   Institution of Civil Engineers
   Association of Consulting Engineers
   Civil Engineering Contractors Association

There are also now minor works and design and construct versions of the ICE Conditions and the New Engineering Contract.

For use on civil engineering works in an international context, there is a family of contracts produced by the Fédération Internationale des Ingénieurs-Conseils, known as the FIDIC contracts, with guidance notes and publications dealing with tendering procedures, insurance matters, risk management, quality management and the settlement of disputes.

Other so-called standard forms of contract have been issued unilaterally by specialist organisations. If used at all, great care must be exercised as the employer's interests may not be protected. It is also worth noting, when working overseas, that some countries have their own standard forms of contract which may need to be considered, although many of these were originally based on the contracts listed above.

Before any form of contract is used, it is prudent to check with the publishers to ensure that you have the latest edition or amendment and that it complies with all current legislation.

Index

(Main entries in bold)